Basic GIS Coordinates
Third Edition

Basic GIS Coordinates
Third Edition

Jan Van Sickle

CRC Press
Taylor & Francis Group
Boca Raton London New York

CRC Press is an imprint of the
Taylor & Francis Group, an **informa** business

CRC Press
Taylor & Francis Group
6000 Broken Sound Parkway NW, Suite 300
Boca Raton, FL 33487-2742

First issued in paperback 2020

ISBN 13: 978-0-367-57351-5 (pbk)
ISBN 13: 978-1-4987-7462-8 (hbk)

Library of Congress Cataloging-in-Publication Data

Names: Van Sickle, Jan, author.
Title: Basic GIS coordinates / Jan Van Sickle.
Description: Third edition. | Boca Raton : Taylor & Francis, 2017. | Includes bibliographical references and index.
Identifiers: LCCN 2016059466 | ISBN 9781498774628 (hardback : alk. paper)
Subjects: LCSH: Grids (Cartography)--Data processing. | Geographic information systems.
Classification: LCC GA116 .V36 2017 | DDC 910.285--dc23
LC record available at https://lccn.loc.gov/2016059466

Visit the Taylor & Francis Web site at
http://www.taylorandfrancis.com

and the CRC Press Web site at
http://www.crcpress.com

Contents

Preface

A fundamental change is coming. Two datums at the foundation of the current geographic information system (GIS) work will be replaced: the North American Datum 1983 and the North American Vertical Datum 1988. This book provides some of the information necessary to cope with that change. It is a book about coordinates.

Coordinates? Press a few buttons on a computer and they are automatically imported, exported, rotated, translated, collated, annotated, and served up in any format you choose with no trouble at all. There is really nothing to it. Why have a book about coordinates?

It is a good question really. Computers are astounding in their ability to make the mathematics behind coordinate manipulation transparent to the user. This book is not much about that sort of mathematics. However, it is about coordinates and coordinate systems. It is about understanding how these systems work and how they sometimes do not work. It is about how points that should be in New Jersey end up in the middle of the Atlantic Ocean even if the computer has done nothing wrong. And that is, I suppose, the answer to the question from my point of view. Computers are currently very good at repetition and very bad at interpretation. People are usually not so good at repetition. We tend to get bored. However, we can be very good indeed at interpretation, that is, if we have the information in our heads to understand what we are interpreting. This book is about providing some of that sort of information on the subject of coordinates and coordinate systems.

Coordinates are critical to GIS, cartography, and surveying. They are their foundation. Points, lines, and polygons form the geometry. Lines define the polygons, points define the lines, and coordinates define the points. Therefore, at bottom, it is the coordinates that tie the real world to their electronic image in the computer.

There are more than 1000 horizontal geodetic datums and over 3000 Cartesian coordinate systems sanctioned by governments around the world and currently in use to describe our planet electronically and on paper.

Author

Jan Van Sickle has many years of experience in GIS, global positioning system (GPS), surveying, mapping, and imagery. He began working with GPS in the early 1980s when he supervised control work using the Macrometer, the first commercial GPS receiver. He created and led the GIS department at Qwest Communications, Denver, Colorado, for the company's 25,000-mile worldwide fiber optic network. He also led the team that built the GIS for natural gas gathering in the Barnett Shale. He has led nationwide seminars based on his three books: *GPS for Land Surveyors, Basic GIS Coordinates,* and *Surveying Solved Problems.* He led the team that collected, processed, and reported control positions for more than 120 cities around the world for the ortho-rectification of satellite imagery now utilized in a global web utility. He managed the creation of the worldwide test and evaluation (T&E) sites for two major earth observation satellites that are used for frequent accuracy assessments. He created an imagery-based system of deriving road centerlines that meet the stringent Advanced Driver Assistance specifications and developed a method of forest inventory to help quantify that depleted resource in Armenia. He assisted the supervision of the first GPS survey of the Grand Canyon for the photogrammetric evaluation of sandbar erosion along the Colorado River. He has performed three-dimensional mapping with terrestrial photogrammetry and light detection and ranging (LiDAR) as well as Building Information Modeling for major buildings in Washington, DC. He was a member of the team of authors for the Geospatial Technology Competency Model for the Department of Labor. He has provided technical assistance in the reconstruction of the geodetic network of Nigeria. He has managed the gravity/magnetic and hyperspectral/multispectral analysis of Borzon VII, a concession block in the South Gobi in Mongolia. He writes the *From Scratch* blog (http://www.sparpointgroup.com/jan-van-sickle) for the SPAR Point Group. He has contributed a chapter on satellite-based surveying technology to the update of the more than 30-year old American Society of Civil Engineers Engineering Surveying Manual of Practice and course work on global navigation satellite systems (GNSS) to the American Association for Geodetic Surveying Geodetic Certification Program. He has been a featured speaker at many conferences, including MAPPS, geospatial information and technology association (GITA), and the Institute of Navigation Annual Meeting. He is a senior lecturer at Pennsylvania State University, State College, Pennsylvania, and teaches at the University of Colorado Denver, Denver, Colorado. He was formerly on the board of Rocky Mountain - American Society of Photogrammetry and Remote Sensing (RM-ASPRS), the vice-chairman of GIS in the Rockies, and the founding chairman of the U.S. West chapter of the Americas Petroleum Survey Group. Sickle earned a PhD in geospatial engineering at the University of Colorado. He has been a licensed professional land surveyor for 36 years and is a member of the Royal Institute of Chartered Surveyors and currently licensed in Colorado, California, Oregon, Texas, North Dakota, West Virginia, and Pennsylvania. His website is www.janvansickle.com.

1

Foundation of a Coordinate System

Uncertainty

Coordinates are slippery devils. A stake driven into the ground holds a clear position, but it is awfully hard for its coordinates to be so certain, even if the figures are precise. For example, a latitude of 40° 25′ 33.504″ N with a longitude of 108° 45′ 55.378″ W appears to be an accurate and unique coordinate. Actually, it could correctly apply to more than one place. An elevation, or height, of 2658.2 m seems unambiguous too, but it is not.

In fact, this latitude, longitude, and height once pinpointed a control point known as Youghall, but not anymore. Oh, Youghall still exists. It is a bronze disk cemented into a drill hole in an outcropping of bedrock on Tanks Peak in the Colorado Rocky Mountains. But its coordinates have not been nearly as stable as the monument. In 1937 the *United States Coast and Geodetic Survey* set Youghall at a latitude of 40° 25′ 33.504″ N and longitude 108° 45′ 55.378″ W. You might think that was that, but in November of 1997 Youghall suddenly got a new coordinate, 40° 25′ 33.39258″ N and 108° 45′ 57.78374″ W, that is more than 56 m, 185 ft, west and 3 m, 11 ft, south of where it started. Its elevation changed too. It was 2658.2 m in 1937. It is 2659.6 m today. It rose to 4½ ft.

Of course, it did no such thing; the station is right where it has always been. The Earth shifted underneath it, well, nearly. It was the *datum* that changed. The 1937 latitude and longitude for Youghall was based on the *North American Datum 1927 (NAD27)*. Sixty years later, in 1997 the basis of the coordinate of Youghall became the *North American Datum 1983 (NAD83)*.

Datums to the Rescue

Coordinates without a specified datum are vague. It means that questions like *Height above what?*, *Where is the origin?*, and *On what surface do they lay?* go unanswered. When that happens coordinates are of really no use. An origin, a starting place, is a necessity for them to be meaningful. Not only must they have an origin, they must be on a clearly defined surface. These foundations constitute the *datum*.

Without a datum, coordinates are like checkers without a checkerboard, you can arrange them, analyze them, and move them around, but absent the framework, you never really know what you have got. In fact, datums very like checkerboards have been in use for a long time. They are generally called *Cartesian*.

René Descartes

Cartesian systems got their name from René Descartes, a mathematician and philosopher. In the world of the seventeenth century he was also known by the Latin name Renatus Cartesius, which might explain why we have a whole category of coordinates known as Cartesian coordinates. Descartes did not really invent the things, despite a story of him watching a fly walk on his ceiling and then tracking the meandering path with this system of coordinates. Long before, around 250 BC or so the Greek mathematician, geographer, and astronomer, Eratosthenes, used a checkerboard-like grid to locate positions on the Earth and even he was not the first. Dicaearchus had come up with the same basic idea before him. Nevertheless, Descartes was probably the first to use graphs to plot and analyze mathematical functions. He set up the rules we use now for his particular version of a coordinate system in two dimensions defined on a flat plane by two axes.

Cartesian Coordinates

Two-dimensional Cartesian coordinates are expressed in ordered pairs. Each element of the coordinate pair is the distance measured across the flat plane parallel with an axis. If the measurement is parallel with the x-axis, it is called the x-coordinate, and if the measurement is parallel with the y-axis, it is called the y-coordinate.

Figure 1.1 shows two axes perpendicular to each other labeled x and y. This labeling is a custom established by Descartes. His idea was to symbolize unknown quantities with letters at the end of the alphabet: x, y, and z. This leaves letters at the beginning that are available for known values. Coordinates became so often used to solve for unknowns, the principle that Cartesian axes have the labels x and y was established. The fancy names for the axes are the *abscissa*, for x, and *ordinate*, for y. Surveyors, cartographers, and mappers call them north and east, but back to the story.

These axes need not be perpendicular with each other. They could intersect at any angle, though they would obviously be of no use if they were parallel. But so much convenience would be lost using anything other than a right angle, it has become the convention. Another convention is the idea that the units along the x-axis are identical with the units along the y-axis, even though there is no theoretical requirement that this be so. Finally, on

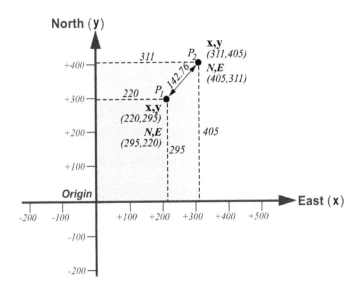

FIGURE 1.1
The Cartesian coordinate system.

the x-axis, any point to the west, that is left, of the origin is negative, and any point to the east, to the right, is positive. Similarly, on the y-axis, any point north of the origin is positive; and south, negative. If these principles are held, then the rules of Euclidean geometry are true and the off-the-shelf CAD and GIS software on your PC have no trouble at all working with these coordinates, a most practical benefit.

For example, the distance between these points can be calculated using the coordinate geometry you learned in high school. The x and y coordinates for the points in the illustration are the origin point, P_1 (220, 295) and point, P_2 (311, 405) therefore where:

$$x_1 = 220$$

$$y_1 = 295$$

and

$$x_2 = 311$$

$$y_2 = 405$$

$$\text{Distance} = \sqrt{\left[(x_1)-(x_2)\right]^2 + \left[(y_1)-(y_2)\right]^2}$$

$$\text{Distance} = \sqrt{\left[(220)-(311)\right]^2 + \left[(295)-(405)\right]^2}$$

$$\text{Distance} = \sqrt{(-91)^2 + (-110)^2}$$

$$\text{Distance} = \sqrt{8,281 + 12,100}$$

$$\text{Distance} = \sqrt{20,381}$$

$$\text{Distance} = 142.76$$

The system works. It is convenient. But it is not very helpful unless it has an attachment to something a bit more real than these unitless numbers, which brings up an important point about datums.

Attachment to the Real World

The beauty of datums is that they are errorless, at least in the abstract. On a datum every point has a unique and accurate coordinate. There is no distortion. There is no ambiguity. For example, the position of any point on the datum can be stated exactly, and it can be accurately transformed into coordinates on another datum with no discrepancy whatsoever. All of these wonderful things are possible only as long as a datum has no connection to anything in the physical world. In that case it is perfectly accurate, and perfectly useless.

But suppose you wished to assign coordinates to objects on the floor of a very real rectangular room. A Cartesian coordinate system could work, if it is fixed to the room with a well-defined orientation. For example, you could put the origin at the southwest corner, stipulate that the walls of the room are oriented in cardinal directions, and use the floor as the reference plane.

With this datum you not only have the advantage that all of the coordinates are positive, but you can define the location of any object on the floor of the room. The coordinate pairs would consist of two distances: (1) the distance east and (2) the distance north from the origin in the corner. As long as everything stays on the floor, you are in business. In this case there is no error in the datum, of course, but errors in the coordinates are inevitable. These errors are due to the less than perfect flatness of the floor, the impossibility of perfect measurement from the origin to any object, the ambiguity of finding the precise center of any of the objects being assigned coordinates, and so on. In short, as soon as you bring in the real world, things get messy.

Cartesian Coordinates and the Earth

Cartesian coordinates then are rectangular, or *orthogonal* if you prefer, defined by perpendicular axes from an origin, along a specifically oriented reference surface. These elements can define a framework, a foundation, and a datum for meaningful coordinates.

As a matter of fact, two-dimensional Cartesian coordinates are an important element in the vast majority of coordinate systems, the *State Plane coordinates* in the United States, the *Universal Transverse Mercator (UTM)* coordinate system, and others. The datums for these coordinate systems are well established. But there are also local Cartesian coordinate systems whose origins are often entirely arbitrary. For example, if surveying, mapping, or other work is done for the construction of a new building, there may be no reason for the coordinates used to have any fixed relation to any other coordinate systems. In that case a local datum may be chosen for the specific project with north and east fairly well defined and the origin moved far to the west and south of the project to ensure that there will be no negative coordinates. Such an arrangement is good for local work, but it does preclude any easy combination of such small independent systems. Large-scale Cartesian datums, on the other hand, are designed to include positions across significant portions of the Earth's surface into one system. Of course, these are also designed to represent our decidedly round planet on the flat Cartesian plane, no easy task (Figure 1.2).

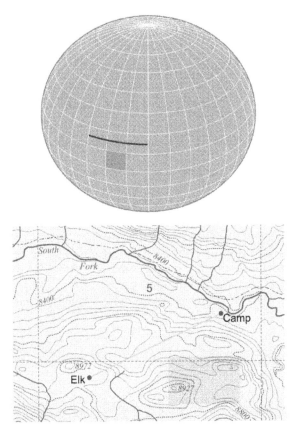

FIGURE 1.2
Distortion increases over long lines and large areas.

But how would a flat Cartesian datum with two axes represent the Earth? There is obviously distortion inherent in the idea. If the planet were flat it would do nicely of course, and across small areas that very approximation, a flat Earth, works reasonably well. That means that even though the inevitable warping involved in representing the Earth on a flat plane cannot be eliminated, it can be kept within well-defined limits as long as the region covered is small and precisely defined. If the area covered becomes too large distortion does defeat it. So the question is, *Why go to all the trouble to work with plane coordinates?* Well, here is a short example.

It is certainly possible to calculate the distance from station Youghall to station Karns using latitude and longitude, also known as *geographic coordinates*, but it is easier for your computer, and for you, to use Cartesian coordinates. Here are the geographic coordinates for these two stations, Youghall at latitude 40° 25′ 33.39258″ N and longitude 108° 45′ 57.78374″ W and Karns at latitude 40° 26′ 06.36758″ N and longitude 108° 45′ 57.56925″ W in the North American Datum 1983 (NAD83). Here are the same two station's positions expressed in Cartesian coordinates.

Youghall

$$\text{Northing} = y_1 = 1{,}414{,}754.47$$

$$\text{Easting} = x_1 = 2{,}090{,}924.62$$

Karns

$$\text{Northing} = y_2 = 1{,}418{,}088.47$$

$$\text{Easting} = x_2 = 2{,}091{,}064.07$$

The Cartesian system used here is called state plane coordinates in *Colorado's North Zone,* and the units are *survey feet,* more about those later. The important point is this; these coordinates are based on a simple two-axis Cartesian system operating across a flat reference plane.

As before the distance between these points using the plane coordinates is easy to calculate.

$$\text{Distance} = \sqrt{(x_1 - x_2)^2 + (y_1 - y_2)^2}$$

$$\text{Distance} = \sqrt{(2{,}090{,}924.62 - 2{,}091{,}064.07)^2 + (1{,}414{,}754.47 - 1{,}418{,}088.47)^2}$$

$$\text{Distance} = \sqrt{(-139.45)^2 + (-3334.00)^2}$$

$$\text{Distance} = \sqrt{(19,445.3025) + (11,115,556.0000)}$$

$$\text{Distance} = \sqrt{11,135,001.30}$$

$$\text{Distance} = 3336.91 \text{ ft}$$

It is 3336.91 ft. The distance between these points calculated from their latitudes and longitudes is slightly different, it is 3337.05 ft. Both of these distances are the result of *inverses*, which means they were calculated between two positions from their coordinates (Figure 1.3).

When you compare the results between the methods you see a difference of about 0.14 ft, a bit more than a tenth of a foot. In other words, the spacing between stations would need to grow more than seven times, to about 4½ miles, before the difference would reach 1 ft. So part of the answer to the

FIGURE 1.3
Youghall to Karns.

question, *Why go to all the trouble to work with plane coordinates?* is this, they are easy to use and the distortion across small areas is not severe.

This rather straightforward idea is behind a good deal of the conversion work done with coordinates. Geographic coordinates are useful but somewhat cumbersome at least for conventional trigonometry. Cartesian coordinates on a flat plane are simple to manipulate but inevitably include distortion. Moving from one to the other it is possible to gain the best of both. The question is *How do you project coordinates from the nearly spherical surface of the Earth to a flat plane?* Well, first you need a good model of the Earth.

The Shape of the Earth

People have been proposing theories about the shape and size of the planet for a long time. Despite the fact that local topography is obvious to an observer standing on the Earth, efforts to grasp the more general nature of the planet's shape and size have been occupying scientists for at least 2,300 years. There have, of course, been long intervening periods of unmitigated nonsense on the subject. Ever since 200 BC when Eratosthenes almost calculated the planet's circumference correctly, geodesy has been getting ever closer to expressing the actual shape of the Earth in numerical terms (Figure 1.4).

At noon the reflection of the midsummer sun was there in the water of a deep well at Syene. The sun was directly overhead. On the same day, measurement of the noon shadow was cast by a pillar at Alexandria. It showed that the sunbeam strikes the earth at an angle of 7.2° off the vertical. Eratosthenes reasoned that the angle between Alexandria and Syene must be 7.2°—one fiftieth of the 360° circle. Syene is 491 miles south of Alexandria and a great circle must therefore be 50 times 491 miles in length—24,550 miles. In fact, the circumference of the earth is around 24,900 miles.

A real breakthrough came in 1687 when Newton suggested that the Earth shape was ellipsoidal in the first edition of his *Principia*. The idea was not entirely without precedent. Years earlier in 1672, astronomer J. Richer found the closer he got to the equator, the more he had to alter the pendulum on his one-second clock. The clock lost about 2½ minutes every day because its pendulum swung more slowly in Cayenne, Guyana, and South America than it did in Paris. In South America he had to shorten the pendulum about 3 mm to bring it back into regulation. When Newton heard about it he speculated that the force of gravity was less in South America than in France. He explained the weaker gravity by the proposition that when it comes to the Earth there is simply more of it around the equator. He wrote, "The Earth is higher under the equator than at the poles, and that by an excess of about 17 miles" (*Philosophiae naturalis principia mathematica*, Book III, Proposition XX). He was pretty close to right; the actual distance is about 13 miles, only about 4 miles less than he thought.

Some supported Newton's idea that the planet bulges around the equator and is flattened at the poles, but others disagreed, for example, the director of

FIGURE 1.4
Eratosthenes data.

the Paris Observatory, Jean Dominique Cassini. Even though he had seen the flattening of the poles of Jupiter in 1666, neither he nor his son Jacques were prepared to accept the same idea when it came to the Earth. And it appeared they had some evidence on their side.

For geometric verification of the Earth model, scientists had employed measurements along a meridian of longitude since the early 1500s. First, using astronomy, they would establish two points on the meridian 1° of latitude apart. Next, they would measure the length of that 1° along the ground. Early attempts assumed a spherical Earth, and the results were used to estimate its radius by simple multiplication. In fact, one of the most accurate of the measurements of this type, begun in 1669 by the French astronomer and priest abbé J. Picard, was used by Newton in formulating his law of gravitation. However, Cassini noted that close analysis of Picard's arc measurement, and others, seemed to show that the length along a meridian through 1° of latitude actually *decreased* as it proceeded northward. If that was true then the Earth was elongated at the poles, not flattened.

The argument was not resolved until Anders Celsius, a famous Swedish physicist on a visit to Paris, suggested two expeditions. One group, led by Moreau de Maupertuis, went to measure a meridian arc along the Tornio River near the Arctic Circle, 66° 20′ N latitude, in Lapland. Another expedition went to what is now Ecuador, to measure a similar arc near the equator, 01° 31′ S latitude. The Tornio expedition reported that 1° along the meridian in Lapland was 57,437.9 *toises*, about 69.6 miles. A *toise* is approximately 6.4 ft. A degree along a meridian near Paris had been measured as 57,060 toises, 69.1 miles. This shortening of the length of the arc was taken as a proof that the Earth is flattened near the poles. Even though the measurements were wrong, the conclusion was correct. Maupertuis published a book on the work in 1738, the king of France gave Celsius a yearly pension of 1,000 livres, and Newton's conjecture was proved right. I wonder which of them was the most pleased.

Since then there have been numerous meridian measurements all over the world, not to mention satellite observations, and it is now settled that the Earth most nearly resembles an oblate spheroid. An oblate spheroid is an ellipsoid of revolution. In other words, it is the solid generated when an ellipse is rotated around its shorter axis and then flattened at its poles. The flattening is only about one part in 300. Still the ellipsoidal model, bulging at the equator and flattened at the poles, is the best representation of the general shape of the Earth. If such a model of the Earth were built with an equatorial diameter of 25 ft the polar diameter would be about 24 ft 11 in., almost indistinguishable from a sphere.

It is on this somewhat ellipsoidal Earth model that latitude and longitude have been used for centuries. The idea of a nearly spherical grid of imaginary intersecting lines has helped people to navigate around the planet for more than a thousand years and is showing no signs of wearing down. It is still a convenient and accurate way of defining positions.

Latitude and Longitude

Latitude and longitude are the two coordinates that represent a position on the surface of the Earth with angles instead of distances. Usually the angles are measured in degrees, but *grads* and *radians* are also used. Depending on the precision required, the degrees, 360° comprising a full circle, can be subdivided into 60 *minutes of arc*, which are themselves divided into 60 *seconds of arc*. In other words, there are 3600 s in a degree. Seconds can be subsequently divided into decimals of seconds. Typically, the arc is dropped from their names, since it is usually obvious that the minutes and seconds are in space rather than time. In any case these subdivisions are symbolized by ° for degrees, ' for minutes, and " for seconds. The system is called *sexagesimal*. A radian is the angle subtended by an arc equal to the radius of a circle. A full circle is 2π radians and a single radian is 57° 17' 44. 8" or 66.66198 grad.

In the European *centesimal* system a full circle is divided into 400 grads. These units are also known as *grades* and *gons*. A radian is the angle subtended by an arc equal to the radius of a circle. A full circle is 2π radians and a single radian is 57° 17' 44. 8".

Lines of latitude and longitude always cross each other at right angles, as do the lines of a Cartesian grid, but latitude and longitude exist on a curved rather than a flat surface. There is imagined to be an infinite number of these lines on the ellipsoidal model of the Earth. In other words any and every place has a line of latitude and a line of longitude passing through it, and it takes both of them to fully define a place. If the vertical distance from the surface of the ellipsoid to the point is then added to the latitude and longitude, you have one type of three-dimensional (3D) coordinate. This distance component is sometimes the elevation above the ellipsoid, also known as the ellipsoidal height. But for the moment the discussion will just be concerned with positions right on the ellipsoidal model of the Earth. There the height component can be set aside for the moment with the assertion that all positions are on the surface of the model.

In mapping, latitude is usually represented by the small Greek letter phi, φ. Longitude is usually represented by the small Greek letter lambda, λ. In both cases the angles originate at a plane that is imagined to intersect the ellipsoid. In both latitude and longitude the planes of origination are intended to include the center of the Earth. Angles of latitude most often originate at the plane of the equator, and angles of longitude originate at the plane through an arbitrarily chosen place, now Greenwich, England. Latitude is an angular measurement of the extent a particular point lays north or south of the equatorial plane measured in degrees, minutes, seconds, and usually decimals of a second. Longitude is also an angle measured in degrees, minutes, seconds, and decimals of a second east and west of the plane through the chosen prime, or zero, position.

Between the Lines

On the Earth any two meridians of longitude, for example, west longitude 89° 00' 00" and west longitude 90° 00' 00", are farthest from each other at the equator but as they proceed north and south to the poles they become closer. In other words, they *converge*. It is interesting to note that the length of a degree of longitude and the length of a degree of latitude are just about the same in the vicinity of the equator. They are both about 60 nautical miles, around 69 miles. But if you imagine going north or south along a line of longitude toward either the North or the South Pole, a degree of longitude becomes progressively shorter. At two-thirds of the distance from the equator to the pole at 60° north and south latitudes, a degree of longitude shrinks to about 55.5 km or 34.6 miles—half the length it had at the equator. And as one proceeds northward or southward a degree of longitude continues to shrink until it fades away to nothing as shown in Figure 1.5.

FIGURE 1.5
Distances across 1°.

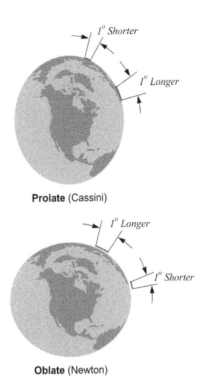

Prolate (Cassini)

Oblate (Newton)

FIGURE 1.6
Prolate and oblate.

On the other hand, lines of latitude do not converge. They are always parallel with one another and the equator. In fact, as one approaches the poles, where a degree of longitude becomes small, a degree of latitude actually grows slightly. This small increase is due to the flattening near the poles mentioned in the discussion of the oblate shape of the planet earlier, see Figure 1.6. The increase in the size of a degree of latitude would not happen if the Earth were a sphere; in that case the length of a degree of latitude would consistently be approximately 110.6 km or 68.7 miles as it is near the equator. However, since the Earth is an oblate spheroid, as Newton predicted, a degree of latitude actually gets a bit longer at the poles. It grows to about 111.7 km or 69.4 miles in that region, which is what all those scientists were trying to measure back in the eighteenth century.

Longitude

Longitude is an angle between two planes. It is a *dihedral angle*. In other words it is an angle measured at the intersection of two planes that are perpendicular to the plane of the equator. In the case of longitude, the first plane passes through the point of interest, the place whose longitude you wish to know, and the second plane passes through an arbitrarily chosen point representing zero longitude. Today, that place is Greenwich, England. The measurement of angles of longitude is imagined to take place where the two planes meet, and that place is the line known as the polar axis. As it happens, that line is also the axis of rotation of the aforementioned ellipsoidal model of the Earth. And where they intersect that ellipsoidal model they create an elliptical line on its surface. This elliptical line is then divided into two meridians at the polar axis. One half becomes a meridian of east longitude, which is labeled E or given a positive (+) values, and the other half a meridian of west longitude, which is labeled W or given a negative (−) value as shown in Figure 1.7.

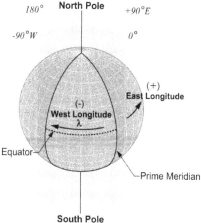

FIGURE 1.7
Longitude.

The zero meridian through Greenwich is called the *prime meridian*. From there, meridians range from +0° to +180° E longitude and −0° to −180° W longitude. Taken together these meridians cover the entire 360° around the Earth. This arrangement was one of the decisions made by consensus of 25 nations in 1884.

The location of the prime meridian is arbitrary. It was formally established at the International Meridian Conference in October 1884 in Washington, DC that it passes through the principal transit instrument, the main telescope, at the Observatory at Greenwich, England, today it is reckoned to be about 102 m east of it. This is due to the deflection of the vertical and there will be a little more about that later in this chapter. There it was decided that there would be a single zero meridian rather than the many used before. There were several other decisions made at the meeting, and among them was the agreement that all longitude would be calculated both east and west from this meridian up to 180°, east longitude is positive and west longitude is negative.

The 180° meridian is a unique longitude, similar to the prime meridian it divides the Eastern Hemisphere from the Western Hemisphere, but it also represents The International Date Line. The calendars west of the line are one day ahead of those east of the line. This division could theoretically occur anywhere on the globe but it is convenient for it to be 180° from Greenwich in a part of the world mostly covered by ocean. Even though the line does not actually follow the meridian exactly, it avoids dividing populated areas; it illustrates the relationship between longitude and time. Since there are 360° of longitude and 24 hours in a day it follows that the Earth must rotate at a rate of 15° per hour. This idea is inseparable from the determination of longitude.

Latitude

Two angles are sufficient to specify any location on the reference ellipsoid representing the Earth. Latitude is an angle between a plane and a line through a point.

Imagine a flat plane intersecting an ellipsoidal model of the Earth. If the plane is coincident or parallel with the equator the result is a *parallel of latitude*. The equator is a unique parallel of latitude that also contains the center of the ellipsoid as shown in Figure 1.8.

The equator is 0° latitude, and the North and South Poles are situated at +90° north and −90° south latitude, respectively. In other words, values for latitude range from a minimum of 0° to a maximum of 90°. The latitudes north of the equator are positive and those to the south are negative.

Lines of latitude are called parallels because they are always parallel to each other as they proceed around the globe. They do not converge as meridians do or cross each other.

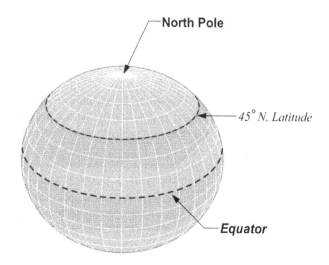

FIGURE 1.8
Parallels of latitude.

Categories of Latitude and Longitude

There are several types of geographic coordinates. For example, there are geocentric and geodetic versions of latitude and longitude.

The geodetic longitude of a point is the angle between the plane of the Greenwich meridian and the plane of the meridian that passes through the point of interest, both planes being perpendicular to the equatorial plane. Since they both have the same zero meridian and the same axis, geodetic longitude and geocentric longitude are equivalent, but when it comes to latitude that is not the case.

It is the ellipsoidal nature of the model of the Earth that contributes to the difference. For example, these are just a few special circumstances on an ellipsoid where a line from a particular position can be both perpendicular to the ellipsoids surface and also pass through the center. Lines from the poles and lines from the equatorial plane can do that but in every other case a line can either be perpendicular to the surface of the ellipsoid or it can pass through the center, but it cannot do both. And there you have the basis for the difference between geocentric and geodetic latitude.

Imagine a line from the point of interest on the ellipsoid to the center of the Earth. The angle that line makes with the equatorial plane is the point's geocentric latitude. On the other hand geodetic latitude is derived from a line that is perpendicular to the ellipsoidal model of the Earth at the point of interest. The angle this line makes with the equatorial plane of that ellipsoid is called geodetic latitude. As you can see geodetic latitude is always just a bit larger than the geocentric latitude except at the poles and the equator, where

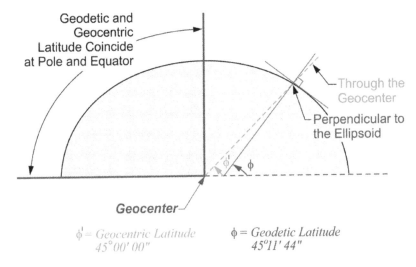

FIGURE 1.9
Geocentric and geodetic latitude.

they are the same. The maximum difference between geodetic and geocentric latitude is about 11′ 44″ and occurs at about 45° (Figure 1.9).

When latitude and longitude are mentioned without a particular qualifier in most cases it is best to presume that the reference is to geodetic latitude and longitude.

The Deflection of the Vertical

Down seems like a pretty straightforward idea. A hanging plumb bob certainly points down. Its string follows the direction of gravity. That is one version of the idea. There are others too.

Imagine an optical surveying instrument set up over a point. If it is centered precisely with a plumb bob and leveled carefully, the plumb line and the line of the level telescope of the instrument are perpendicular to each other. In other words, the level line, the horizon of the instrument, is perpendicular to gravity. Using an instrument so oriented it is possible to determine the latitude and longitude of the point. Measuring the altitude of a circumpolar star is one good method of finding the latitude. The measured altitude would be relative to the horizontal level line of the instrument. Latitude found this way is called *astronomic latitude.*

One might expect that this astronomic latitude would be the same as the geocentric latitude of the point, but they are different. The difference is due to the fact that a plumb line coincides with the direction of gravity, it does not point to the center of the Earth where the line is used to derive geocentric latitude originates.

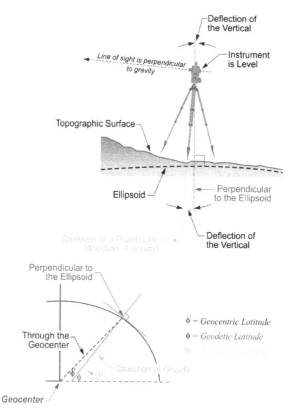

FIGURE 1.10
Geocentric, geodetic, and astronomic latitude.

Astronomic latitude also differs from the most widely used version of latitude, geodetic. The line from which geodetic latitude is determined is perpendicular with the surface of the ellipsoidal model of the Earth. That does not match a plumb line either. In other words, there are three different versions of down and each with its own latitude. For geocentric latitude, down is along a line to the center of the Earth. For geodetic latitude, down is along a line perpendicular to the ellipsoidal model of the Earth. For astronomic latitude, down is along a line in the direction of gravity. And more often than not, these are three completely different lines (Figure 1.10).

Each can be extended upward too, toward the zenith, and there are small angles between them. The angle between the vertical extension of a plumb line and the vertical extension of a line perpendicular to the ellipsoid is called the *deflection of the vertical*. It sounds better than the difference in down. This deflection of the vertical defines the actual angular difference between the astronomic latitude and longitude of a point and its geodetic latitude and longitude. Latitude and longitude because, even though the discussion has so far been limited to latitude, the deflection of the vertical usually has both a north–south and an east–west component.

It is interesting to note that optical instrument set up so carefully over a point on the Earth cannot be used to measure geodetic latitude and longitude directly because they are not relative to the actual Earth, but rather a model of it. Gravity does not even come into the ellipsoidal version of down. On the model of the Earth down is a line perpendicular to the ellipsoidal surface at a particular point. On the real Earth down is the direction of gravity at the point. They are most often not the same thing. It is quite impossible to actually set up an instrument on the ellipsoid because it is imaginary. On the other hand astronomic observations, observing stars and planets, for the measurement of latitude and longitude has a very long history indeed. And yet the most commonly used coordinates today are not astronomic latitudes and longitudes, they are geodetic latitudes and longitudes. So conversion from astronomic latitude and longitude to geodetic latitude and longitude has a long history as well.

Until the advent of GPS/GNSS/GNSS geodetic latitudes and longitudes was derived by postobservation calculation from astronomic observations. In a sense that is still true, the change is a modern GPS/GNSS/GNSS receiver can display the geodetic latitude and longitude of a point to the user immediately because the calculations can be completed with incredible speed in the microcomputer in the receiver itself.

This is one of the many changes that have resulted from satellite positioning, navigation, and timing. As more constellations of such satellites become available the term GPS/GNSS by itself is no longer adequate. All are included in the name *Global Navigation Satellite System* (*GNSS*). In this book the abbreviation GPS/GNSS/GNSS will be used to represent all the systems.

Nevertheless, a fundamental fact remains unchanged, the instruments by which latitudes and longitudes are measured are oriented to gravity, the ellipsoidal model on which geodetic latitudes and longitudes are determined is not. And that is just as true for the antenna of a GPS/GNSS receiver, an optical surveying instrument, and a camera in an airplane taking aerial photography, or even the GPS/GNSS satellites themselves.

As a matter of illustration of the effect of the deflection of the vertical on latitude and longitude here are station Youghall's astronomical latitude and longitude labeled with capital phi (Φ) and capital lambda (Λ), the standard Greek letters commonly used to differentiate them from geodetic latitude and longitude:

$$\Phi = 40° \ 25' \ 36.28'' \ N$$

$$\Lambda = 108° \ 46' \ 00.08'' \ W$$

Now the deflection of the vertical can be used to convert these coordinates to a geodetic latitude and longitude. Unfortunately the small angle is not usually conveniently arranged. It would be helpful if the angle between the direction of gravity and the perpendicular to the ellipsoid would follow just one cardinal direction, north–south or east–west, but it does not. For example, if the angle observed from above Youghall was oriented north or south along the meridian, then it would affect only the latitude, not the longitude and would be very easy to apply. But that is not the case. The two *normals* that are the perpendicular lines that constitute the deflection of the vertical at a point are usually neither north–south nor east–west of each other. Looking down on a point one could imagine that the angle they create between them stands in one of the four quadrants: northeast, southeast, southwest, or northwest. Therefore, in order to express its true nature it is broken down into two components, one north–south and the other east–west. There are almost always some of both. The north–south component is known by the Greek letter xi (ξ). It is positive (+) to the north and negative (−) to the south. The east–west component is known by the Greek letter eta (η). It is positive (+) to the east and negative (−) to the west (Figure 1.11).

For example, the components of the deflection of the vertical at Youghall are

$$\text{North–south} = \text{xi} = \xi = +6.18''$$

$$\text{East–west} = \text{eta} = \eta = -0.62''$$

In other words, if an observer held a plumb bob directly over the monument at Youghall the upper end of the string would be 2.89 arc seconds north and 1.75 arc seconds east of the line that is perpendicular to the ellipsoid.

The geodetic latitude and longitude can be computed from the astronomic latitude and longitude given above using the following formulas:

$$\phi = \Phi - \xi$$

$$\phi = 40° \, 25' \, 39.\,57'' - (+6.18'')$$

$$\phi = 40° \, 25' \, 33.\,39''$$

$$\lambda = \Lambda - \eta / \cos \varphi$$

$$\lambda = 108° \, 45' \, 56.\,97'' - (-0.62'') / \cos 40° \, 25' \, 33.\,39''$$

$$\lambda = 108° \, 45' \, 56.\,97'' - (-0.62'') / 0.7612447$$

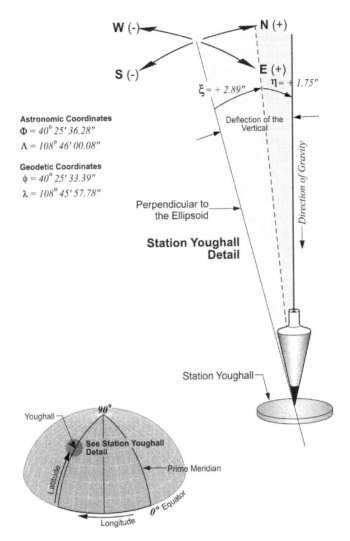

FIGURE 1.11
Two components of the deflection of the vertical at station Youghall.

$$\lambda = 108° \ 45' \ 56.97'' - (-0.81'')$$

$$\lambda = 108° \ 45' \ 57.78''$$

where:
  φ is the geodetic latitude
  λ is the geodetic longitude
  Φ is the astronomical latitude
  Λ is the astronomical longitude

And the components of the deflection of the vertical are

$$\text{North--south} = xi = \xi$$

$$\text{East--west} = eta = \eta$$

Directions

Azimuths

An *azimuth* is one of the ways to define the direction from point-to-point on the ellipsoidal model of the Earth, on Cartesian datums, and others. On some Cartesian datums an azimuth is called a *grid* azimuth, referring to the rectangular grid on which a Cartesian system is built. Grid azimuths are defined by a horizontal angle measured clockwise from north.

Azimuths can be either measured clockwise from north through a full 360° or measured +180° in a clockwise direction from north and −180° in a counterclockwise direction from north. Bearings are different.

Bearings

Bearings, another method of describing directions, are always acute angles measured from 0° at either north or south to 90° to either the west or the east. They are measured both clockwise and counterclockwise. They are expressed from 0° to 90° from north in two of the four quadrants, the northeast, 1, and northwest, 4. Bearings are also expressed from 0° to 90° from south in the two remaining quadrants, the southeast, 2, and southwest, 3 as shown in Figure 1.12.

In other words, bearings use four quadrants of 90° each. A bearing of N 45° 15′ 35″ E is an angle measured in a clockwise direction 45° 15′ 35″ from north toward the east. A bearing of N 21° 44′ 52″ W is an angle measured in a counterclockwise direction 21° 44′ 52″ toward west from north. The same ideas work for southwest bearings measured clockwise from south and southeast bearing measured counterclockwise from south.

Directions—azimuths and bearings—are indispensable. They can be derived from coordinates with an inverse calculation. If the coordinates of two points inversed are geodetic then the azimuth or bearing derived from them is also geodetic, if the coordinates are astronomic then the direction will be astronomic, and so on. If the coordinates from which a direction is calculated are *grid coordinates*, the resulting azimuth will be a grid azimuth, and the resulting bearing will be a grid bearing.

Both bearings and azimuths in the flat Cartesian system assume the direction to north is always parallel with the y-axis, the north–south axis. On a Cartesian datum there is no consideration for convergence of *meridional,*

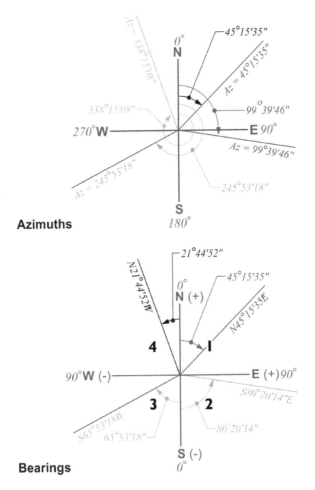

FIGURE 1.12
Azimuths and bearings.

north–south, lines. One result of the lack of convergence is the bearing or azimuth at one end of a line and is always exactly 180° different from the bearing or azimuth at the other end of the same line. But if the datum is on the curved ellipsoidal model of the Earth directions do not quite work that way. For example, consider the difference between an astronomic azimuth and a geodetic azimuth.

Astronomic and Geodetic Directions

If it were possible to point an instrument to the exact position of the north celestial pole, a horizontal angle turned from there to an observed object on the Earth would be the astronomic azimuth to that object from the instrument. But it is rather difficult to measure an astronomic azimuth that way because there is nothing to point to at the celestial North Pole but a lot of sky. Polaris, the North Star, appears to follow an elliptical path around the celestial North Pole, which is the northward prolongation of the Earth's axis. Even so, Polaris and several other celestial bodies, for that matter, serve as good references for the measurement of astronomic azimuths, albeit with a bit of calculation. Still optical instruments used to measure astronomic azimuths must be oriented to gravity, and it is usual for the azimuths derived from celestial observations to be converted to geodetic azimuths. Geodetic azimuths are the native form of azimuths on an ellipsoid. So, as it was with the conversion of astronomic latitudes and longitudes to geodetic coordinates, the deflection of the vertical is applied to convert astronomic azimuths to geodetic azimuths.

For example, the astronomic azimuth between two stations, from Youghall to Karns, is 00° 17′ 06.67″. Given this information, the geodetic latitude of Youghall, 40° 25′ 33.39″ N and the east–west component of the deflection of the vertical, eta = η = +1.75, it is possible to calculate the geodetic azimuth from Youghall and Karns using the following formula:

$$\alpha = A - \eta \tan \phi$$

$$\alpha = 00° \, 17′ \, 06.\,67″ - (+1.75″) \tan 40° \, 25′ \, 33.\,39″$$

$$\alpha = 00° \, 17′ \, 06.\,67″ - (+1.75″) \, 0.851847724$$

$$\alpha = 00° \, 17′ \, 06.\,67″ - 1.49$$

$$\alpha = 00° \, 17′ \, 05.\,18″$$

where:
  α is the geodetic azimuth
  A is the astronomical azimuth
  η is the east–west component of the deflection of the vertical
  ϕ is the geodetic latitude

But as always, there is another way to calculate the difference between an astronomic azimuth and a geodetic azimuth. Here is the formula and a calculation using the data at Youghall:

$$\Phi = 40° \, 25′ \, 36.\,28″ \, N$$

$$\Lambda = 108° \, 46′ \, 00.\,08″ \, W$$

$$\phi = 40° \, 25′ \, 33.\,39″ \, N$$

$$\lambda = 108° \, 45′ \, 57.\,78″ \, W$$

$$\alpha_A - \alpha_G = +(\Lambda - \lambda) \sin \phi$$

$$\alpha_A - \alpha_G = +(108° \ 46' \ 00. \ 08'' - 108° \ 45' \ 57. \ 78'') \sin 40° \ 25' \ 33.39''$$

$$\alpha_A - \alpha_G = + (2.30'') \sin 40° \ 25' \ 33.39''$$

$$\alpha_A - \alpha_G = +(2.30'') \ 0.64846$$

$$\alpha_A - \alpha_G = +1.49''$$

where:
  α_A is the astronomic azimuth
  α_G is the geodetic azimuth
  Λ is the astronomic longitude
  λ is the geodetic longitude
  ϕ is the geodetic latitude

Even without specific knowledge of the components of the deflection of the vertical it is possible to calculate the difference between an astronomic azimuth and a geodetic azimuth. The required information is in the coordinates of the point of interest. Knowing the astronomic longitude and the geodetic latitude and longitude of the position is all that is needed. This method of deriving a geodetic azimuth from an astronomic observation is convenient for surveyors to use to derive the *LaPlace correction,* which is the name given to the right side of the above equation.

North

The reference for directions is north, and each category refers to a different north. Geodetic north differs from astronomic north, which differs from grid north, which differs from magnetic north. The differences between the geodetic azimuths and astronomic azimuths are a few seconds of arc at a given point. Variations between these two are small indeed compared to those found between grid azimuths and magnetic azimuths. For example, while there might be a few seconds between astronomic north and geodetic north, there is usually a difference of several degrees between geodetic north and magnetic north.

Magnetic North

Magnetic north is used throughout the world as the basis for magnetic directions in both the Northern and the Southern Hemispheres, but it will not hold still. The position of the magnetic North Pole is somewhere around

86°N latitude, and 170°W longitude, a long way from the geographic North Pole. To make matters even more interesting the magnetic North Pole is moving at a rate of about 40 miles per year, faster than it used to. In fact, it has moved more than 600 miles since the early nineteenth century.

The Earth's magnetic field is variable. For example, if the needle of a compass at a particular place points 15° west of geodetic north there is said to be a west *declination* of 15°. At the same place 20 years later that declination may have grown to 16° west of geodetic north. This kind of movement is called *secular variation*. It is a change that occurs over long periods and is probably caused by convection in the material at the Earth's core. Declination is one of the two major categories of magnetic variation. The other magnetic variation is called daily or *diurnal* variation.

Daily variation is probably due to the affect of the solar wind on the Earth's magnetic field. Diurnal variation is also known as solar daily variation. As the Earth rotates, a particular place alternately moves toward and away from the constant stream of ionized particles from the Sun. Therefore, it is understandable that the daily variation swings from one side of the mean declination to the other over the course of a day. For example, if the mean declination at a place were 15° west of geodetic north, it might be 14.9° at 8 a.m., 15.0° at 10 a.m., 15.6° at 1 p.m., and again 15.0° at sundown. Such a diurnal variation would be somewhat typical, but in high latitudes or during high solar activity it can grow as large as 9°.

Grid North

The position of magnetic north is governed by natural forces, but grid north is entirely artificial. In Cartesian coordinate systems, whether known as *State Plane, Universal Transverse Mercator*, a local assumed system, or any other, the direction to north is established by choosing one meridian of longitude. The meridian that is chosen is usually in the middle of the area, the *zone* that is covered by the coordinate system. That is why it is frequently known as the *central meridian*. Thereafter, throughout the system, at all points, grid north is along a line parallel with that central meridian. This arrangement purposely ignores the fact that a different meridian passes through each of the points and all the meridians inevitably converge with one another on the curved Earth. Actually, grid north and geodetic north only agree at points along the central meridian of a Cartesian coordinate system, everywhere else in the coordinate system there is an angular difference between the two directions. That angular difference is known as the *convergence*. East of the central meridian grid north is east of geodetic north and the convergence is positive. West of the central meridian grid north is west of geodetic north and the convergence is negative. The approximate grid azimuth of a line is its geodetic azimuth minus the convergence. Therefore it follows that east of the central meridian the grid azimuth of a line is smaller than its

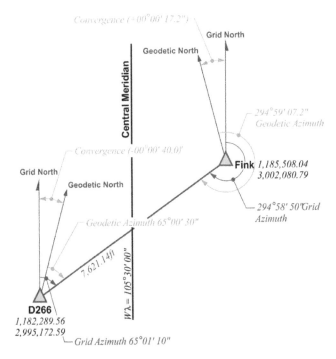

FIGURE 1.13
Approximate grid azimuth = geodetic azimuth − convergence.

geodetic azimuth. West of the central meridian the grid azimuth of a line is larger than the geodetic azimuth as shown in Figure 1.13.

Polar Coordinates

There is another way of looking at a direction. It can be one component of a coordinate.

A procedure familiar to surveyors using optical instruments involves the occupation of a station with an established coordinate. A back sighting is taken either on another station with a coordinate on the same datum or some other reference such as Polaris. With two known positions, the occupied and the sighted, a beginning azimuth or bearing is calculated. Next, a new station is sighted ahead, foresighted, on which a new coordinate will be established. The angle is measured from the back sight to the fore sight, fixing the azimuth or bearing from the occupied station to the new station. A distance is measured to the new station. And this direction and distance together can actually be considered the coordinate of the new station. They constitute

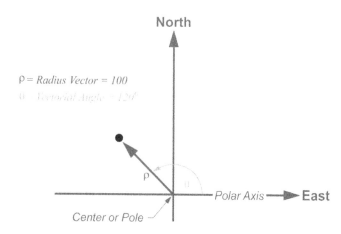

FIGURE 1.14
A polar coordinate (mathematical convention).

what is known as a *polar coordinate*. In surveying, polar coordinates are very often a first step toward calculating coordinates in other systems.

There are coordinates that are all distances—Cartesian coordinates, for example. There are coordinates that are all angles—latitude and longitude, for example. Then there are coordinates that are an angle and a distance—polar coordinates, as shown in Figure 1.14.

A polar coordinate defines a position with an angle and distance. As in a Cartesian coordinate system they are reckoned from an origin, which in this case is also known as the *center* or the *pole*. The angle used to define the direction is measured from the *polar axis*, which is a fixed line pointing to the east, in the configuration used by mathematicians. It is notable that many disciplines presume east as the reference line for directions, computer-aided drafting utilities, for example. Mappers, cartographers, and surveyors tend to use north as the reference for directions in polar coordinates.

In the typical format for recording polar coordinates the Greek letter rho, ρ indicates the length of the *radius vector* that is the line from the origin to the point of interest. The angle from the polar axis to the radius vector is represented by the Greek letter theta, θ and is called the *vectorial angle*, the *central angle*, or the *polar angle*. These values ρ and θ are given in ordered pairs, similar to Cartesian coordinates. The length of the radius vector is first and the vectorial angle second—for example (100,220°).

There is a significant difference between Cartesian coordinates and polar coordinates. In an established datum using Cartesian coordinates one and only one ordered pair can represent a particular position. Any change in either the northing or the easting and the coordinate represents a completely different point. However, in the mathematician's polar coordinates the same position might be represented in many different ways, with many different ordered pairs of ρ and θ standing for the very same point.

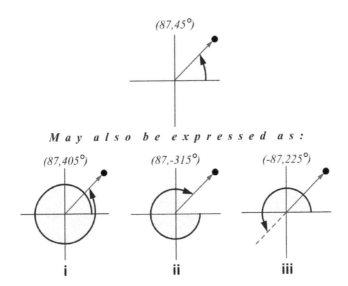

FIGURE 1.15
Four ways of noting one position.

For example, (87,45°) can just as correctly be written (87,405°) as illustrated in (i) in Figure 1.15. Here the vectorial angle swings through 360° and continues past the pole at another 45°. It could also be written as (87,–315) as illustrated in (ii) in Figure 1.13. When θ has a clockwise rotation from the polar axis in this arrangement it is negative. Another possibility is a positive or counterclockwise rotation from the polar axis to a point 180° from the origin and the radius vector is negative (–87,225°). The negative radius vector indicates that it proceeds out from the origin in the opposite direction from the end of the vectorial angle as shown in Figure 1.13.

In other words there are several ways to represent the same point in polar coordinates. This is not the case for rectangular coordinates nor is it the case for the polar coordinate system as used in surveying, mapping, and cartography.

In mapping and cartography directions are consistently measured from north, and the polar axis points toward north as shown in Figure 1.16.

In the mathematical arrangement of polar coordinates a counterclockwise vectorial angle θ is positive and a clockwise rotation is negative. In the surveying, mapping and cartography arrangement of polar coordinates the opposite is true. A counterclockwise rotate is negative and clockwise is positive. The angle may be measured in degrees, radians or grads, but if it is clockwise, it is positive.

In the mathematical arrangement the radius vector can be positive or negative. If the point P lies in the same direction as the vectorial angle, it is considered positive. If the point P lies in the opposite direction back through the origin, the radius vector is considered negative. In the surveying, mapping,

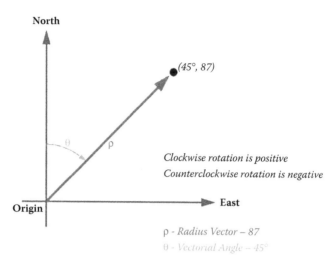

FIGURE 1.16
A polar coordinate (mapping and surveying convention).

and cartography arrangement of polar coordinates, the radius vector always points out from the origin and is always positive.

Summary

Positions in three-dimensional space can be expressed in both Cartesian coordinates and polar coordinates by the addition of a third axes, the z-axis. The z-axis is perpendicular to the plane described by the x-axis and the y-axis. The addition of a third distance in the Cartesian system, or the addition of a second angle in the polar coordinate system completes the three-dimensional coordinates of a point.

The letters φ' and λ' represent the two angles in Figure 1.17. If the origin of the axes are placed at the center of an oblate ellipsoid of revolution the result is a substantially correct model of the Earth from which three-dimensional polar coordinates can be derived. Of course, there is a third element to the polar coordinate here represented by rho, ρ. However, if every position coordinated in the system is always understood to be on the surface of the Earth, or a model of the Earth, this radius vector can be dropped from the coordinate without creating ambiguity. And that is conventionally done, so one is left with the idea that each latitude and longitude comprises a three-dimensional polar coordinate with only two angular parts. In Figure 1.17 they are a geocentric latitude and longitude.

If the three-dimensional polar coordinates represent points on the actual surface of the Earth, the irregularity of the planet's surface presents problems. If they represent points on an ellipsoidal model of the Earth they are on

Polar Coordinate

Cartesian Coordinate

FIGURE 1.17
Three-dimensional polar and Cartesian coordinates.

a regular surface, but that surface does not stand at a constant radial distance from the center of the figure. There is also a difficulty regarding the origin. If the intersection of the axes is at the geocenter, one can derive geocentric latitude and longitude directly. However, the vector perpendicular to the surface of the ellipsoid that represents the element of a geodetic latitude and longitude of a point does not pass through the geocenter, unless the point is at a pole or on the equator. For these reasons and others, it is often convenient to bring in the coordinate system that was presented at the beginning of this chapter, the Cartesian coordinate system. However, this time it is used in its three-dimensional form as shown in Figure 1.17.

A three-dimensional Cartesian coordinate system can be built with its origin at the center of mass of the Earth. The third coordinate, the z-coordinate, is added to the x- and y-coordinates that are both in the plane of the equator. This system can be and is used to describe points on the surface of an ellipsoidal model of the Earth, on the actual surface of the Earth, or satellites orbiting the Earth. This system is sometimes known as the *Earth-Centered Earth-Fixed, ECEF,* coordinate system, more about that will be discussed in Chapter 2.

In Figure 1.18 the relationship between the three-dimensional Cartesian coordinates of two points P_1 and P_2, and their three-dimensional polar coordinates are illustrated on a reference ellipsoid. Under the circumstances the polar coordinates are geocentric latitudes and longitudes. The basic relationship between

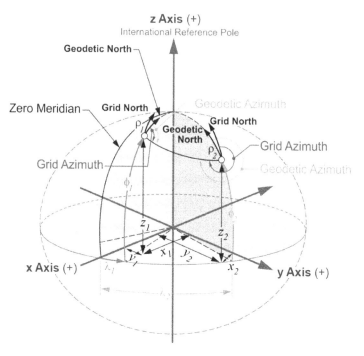

FIGURE 1.18
A few fundamentals.

the geodetic and grid azimuths is also shown in the figure. These elements outline a few of the fundamental ideas involved in commonly used coordinate systems on the Earth. Subsequent chapters will expand on these basics.

Exercises

1. Which of the expressions below is equal to 243.1252326°?
   a. 270.1391473 grads
   b. 218.8127093 grads
   c. 4.24115 radians
   d. 243° 08′ 30″

Explanation:

Since there are 400 grads in a full circle of 360°, each grad is equal to 9/10ths of a degree.

$$\frac{360°}{400\,\text{grads}} = 0.90 \text{ grads in a degree}$$

Therefore, it follows that 243.1252326° divided by 0.90 will yield the measurement in grads.

$$\frac{243.1252326°}{0.90\,\text{grads in a degree}} = 270.1391473\,\text{grads}$$

The decimal degrees converted to degrees, minutes, and seconds is 243° 07′ 30.838″, and the same value can be expressed in radians as 4.24336 radians.

Answer is (a)

2. Which of the following statements about two-dimensional Cartesian coordinates is incorrect?

   a. Universal Transverse Mercator coordinates are Cartesian coordinates.
   b. Most Cartesian coordinate systems are designed to place all coordinates in the first quadrant.
   c. Cartesian coordinates derived from positions on the Earth always include distortion.
   d. Directions in Cartesian coordinate systems are always reckoned from north.

Explanation:

Universal Transverse Mercator coordinates, State Plane coordinates, and many other commonly used systems are indeed two-dimensional and Cartesian. It is desirable and typical to design a flat Cartesian coordinate system to ensure that all coordinates are positive, in other words, all in the first quadrant. It is not possible to represent the curved Earth on a two-dimensional flat surface without some distortion. Many disciplines use Cartesian coordinates and many do not measure directions from the north, or y-axis.

Answer is (d)

3. What is the clockwise angle between the bearings N 27° 32′ 01.34″ W and S 15° 51′ 06.12″ E?

   a. 191° 40′ 55.22″

    b. 168° 19′ 04.78″

    c. 43° 23′ 07.46″

    d. 181° 04′ 22.41″

Explanation:

There is 180° of angle turned clockwise from N 27° 32′ 01.34″ W to S 27° 32′ 01.34″ E. The remaining clockwise angle between the bearings can be found by subtracting 15° 51′ 06.12″ from 27° 32′ 01.34:

$$\begin{array}{r} 27° \ 32' \ 01.34'' \\ -15° \ 51' \ 06.12'' \\ \hline 11° \ 40' \ 55.22'' \end{array}$$

The result, added to 180° is the clockwise angle between N 27° 32′ 01.34″ W and S 15° 51′ 06.12″ E.

$$\begin{array}{r} 180° \ 00' \ 00.00'' \\ +11° \ 40' \ 55.22'' \\ \hline 191° \ 40' \ 55.22'' \end{array}$$

Answer is (a)

4. It is frequently necessary to convert a latitude or longitude that is given in degrees, and decimals of degrees to degrees–minute–seconds and vice versa. Which of the values below correctly reflects the latitude and longitude of Youghall in degrees and decimals of degrees? Its coordinate in degrees, minutes, and seconds is 40° 25′ 33.3926″ N and 108° 45′ 57.783″ W.

    a. +40.1512021, −108.2720802

    b. +40.4259424, −108.7660508

    c. +40.42583926, −108.7658783

    d. +40.15123926, −108.2720783

Explanation:

40° 25′ 33.3926″ N latitude is converted to degrees and decimal degrees by

$$+(40° + 25 \ \text{min} * (1°/60 \ \text{min}) + 33.3926 \ \text{s} * (1 \ \text{min}/60 \ \text{s})(1°/60 \ \text{min}))$$

$$= +40.4259424$$

108° 45′ 57.783″ W longitude is converted to degrees and decimal degrees by

$$-(108° + 45 \text{ min} * (1°/60 \text{ min}) + 57.783 \text{ s} * (1 \text{ min}/60 \text{ s}) * (1°/60 \text{ min}))$$

$$= -108.7660508$$

Answer is (b)

5. Which of the following correctly describes a characteristic of both Cartesian coordinates and polar coordinates share?
   a. Each point has only one unique coordinate pair
   b. Coordinates are expressed in ordered pairs
   c. Angles are measured clockwise from north in degrees, minutes, and seconds
   d. Coordinates are always positive

Explanation:

In the system of Cartesian coordinates each point has a single unique coordinate, any change in the northing, y, or easting, x, and the coordinate refers to another point entirely. In the system of polar coordinates the vectorial angle, ρ and the radius vector, θ can be expressed in a number of ways and still describe the same point. A direction, expressed in an angle, may comprise half of a polar coordinate whether it is measured in grads, radians, decimal degrees or degrees, minutes, and seconds. While it is desirable to place the origin of a Cartesian coordinate system far to the southwest of the area of interest and thereby keep all of the coordinates positive, it is not always the arrangement. Negative coordinates are valid in both polar and Cartesian systems. However, in both systems coordinates are given in ordered pairs.

Answer is (b)

6. As one proceeds northward from the equator, which of the following does not happen?
   a. Meridians converge
   b. Latitudinal lines are parallel
   c. The force of gravity increases
   d. The distance represented by a degree of latitude gets shorter

Explanation:

As Richer discovered that the force of gravity is actually more in the higher latitudes and decreases as one travels southward. As Newton predicted that the Earth is thicker at the Equator. If the Earth were a sphere, meridians would still converge, latitudinal lines would still be parallel and the distance represented by degree of latitude would be constant. However, the Earth is an oblate spheroid and the length of a degree of latitude actually increases slightly as one proceeds northward due to the flattening. It is this fact that originally convinced scientists that Newton's model for the Earth was correct.

Answer is (d)

7. Presuming that all numbers to the right of the decimal are significant, which value below is nearest to the precision of the following coordinates?

$$\phi = 60° \ 14' \ 15.3278'' \ N$$

$$\lambda = 149° \ 54' \ 11.1457'' \ W$$

a. ±10.0 ft
b. ±1.0 ft
c. ±0.10 ft
d. ±0.01 ft

Explanation:

At approximately 60°N latitude the length of a degree of longitude is approximately 34.5 miles. Since there are 3600 s of latitude and longitude in 1°, an estimation of the length on a second of longitude at that latitude can be had by dividing 34.5 miles by 3600. It might be best to convert 34.5 miles to feet first.

$$34.5 \text{ miles} \times 5280 \text{ ft} = 182{,}160 \text{ ft}$$

$$\frac{182{,}160 \text{ ft}}{3600} = 50.6 \text{ ft}$$

The longitude is shown to one ten-thousandth of a degree, so an estimation of the accuracy of the expression can be made by dividing the approximate length of one second by 10,000.

$$\frac{50.6 \text{ ft}}{10{,}000} = 0.005 \text{ ft}$$

The same sort of approximation can be done using the latitude. The length of a degree of latitude is approximately 69 miles. To estimate the length of a second of latitude, divide 69 miles, in feet, by 3600.

$$69 \text{ miles} \times 5280 \text{ ft} = 316{,}800 \text{ ft}$$

$$\frac{316{,}800 \text{ ft}}{3600} = 101.2 \text{ ft}$$

Then, to estimate the length of a ten-thousandth of a second, divide by 10,000.

$$\frac{101.2 \text{ ft}}{10{,}000} = 0.01 \text{ ft}$$

Answer is (d)

8. Which of the following statements about latitude is not true?
   a. The geocentric latitude of a point is usually smaller than the geodetic latitude of the same point.
   b. Geodetic latitude is not derived from direct measurement with optical instruments.
   c. The astronomic latitude is usually quite close to the geodetic latitude of the same point.
   d. The geodetic latitude of a point remains constant despite datum shifts.

Explanation:

It is true that the geocentric latitude of a point is usually smaller than the geodetic latitude of the same point. Instruments set on the Earth are not oriented to the ellipsoid, they are oriented to gravity, and since geodetic latitude is on the ellipsoid it is not feasible to measure it directly. Astronomic latitude departs from geodetic latitude per the deflection of the vertical, usually a very small amount. Further, since the datum for geodetic latitude is an ellipsoidal model of the Earth, shifts in that datum change the latitude, as was illustrated at the beginning of the chapter.

Answer is (d)

9. Which of the following statements concerning the deflection of the vertical is correct?

a. The deflection of the vertical is composed of the north–south and east–west components of the difference between the geodetic and geocentric latitude and longitude of a point.

b. The deflection of the vertical at a point can be derived from astronomic observations alone.

c. The deflection of the vertical at a point can be derived from geodetic coordinates alone.

d. The deflection of the vertical is used in the conversion of astronomic coordinates and astronomic directions to their geodetic counterparts and vice versa.

Explanation:

The deflection of the vertical is pertinent to the differences between astronomic and geodetic coordinates and directions. It has no part to play regarding geocentric information. Calculation of the deflection of the vertical requires both astronomic and geodetic information. It cannot be revealed without some aspect of both.

Answer is (d)

10. Given the astronomical coordinates,

$$\Phi = 39° \, 59' \, 38.66'' \, N$$

$$\Lambda = 104° \, 59' \, 51.66'' \, W$$

the geodetic coordinates of the same point

$$\phi = 39° \, 59' \, 36.54'' \, N$$

$$\lambda = 104° \, 59' \, 38.24'' \, W$$

and an astronomic azimuth of 352° 21′ 14.8″, What is the corresponding geodetic azimuth?

a. 352° 21′ 06.2″

b. 352° 21′ 12.7″

c. 352° 21′ 01.4″

d. 352° 21′ 23.4″

Explanation:

$$\alpha_A - \alpha_G = +(\Lambda - \lambda) \sin \phi$$

$$\alpha_A - \alpha_G = +(104° \, 59' \, 51.66'' - 104° \, 59' \, 38.24'') \sin 39° \, 59' \, 36.54''$$

$$\alpha_A - \alpha_G = +(13.42'') \sin 39° 59' 36.54$$

$$\alpha_A - \alpha_G = +(13.42'') 0.6427$$

$$\alpha_A - \alpha_G = +8.6''$$

where:
α_A is the astronomic azimuth
α_G is the geodetic azimuth
Λ is the astronomic longitude
λ is the geodetic longitude
ϕ is the geodetic latitude
Therefore, given the astronomic azimuth $\alpha_A = 352° 21' 14.8''$:

$$\alpha_A - \alpha_G = +8.6''$$

$$352° 21' 14.8'' - \alpha_G = +8.6''$$

$$352° 21' 14.8'' - 8.6'' = \alpha_G$$

$$352° 21' 06.2'' = \alpha_G$$

Answer is (a)

11. Which of the following statements correctly describes a difference between polar coordinates expressed in the mathematical convention and those expressed in the mapping and surveying convention?

   a. The vectorial angle is measured from east in the mapping and surveying convention but not in the mathematical convention.
   b. The radius vector is symbolized by ρ in the mathematical convention but not in the mapping and surveying convention.
   c. The vectorial angle is measured from north in the mapping and surveying convention but not in the mathematical convention.
   d. The radius vector is measured from the origin in the mapping and surveying convention but not in the mathematical convention.

Explanation:

In mapping, surveying, and cartography directions are consistently measured from north, and the polar axis points north as shown in Figure 1.16.

Answer is (c)

2

Building a Coordinate System

The actual surface of the Earth is not very cooperative. It is bumpy. There is not one nice smooth figure that will fit it perfectly. It does resemble an ellipsoid somewhat, but an ellipsoid that fits Europe may not work for North America as well. And one applied to North America may not be the best for other parts of the planet. That is why, in the past, several ellipsoids were invented to model the Earth. There are about 50 or so still in regular use for various regions of the planet. They have been, and to a large degree still are, the foundation of coordinate systems around the world, but things have changed. The changes are largely a consequence of the advancements in the accuracy and nature of measurement. We also have a much better idea of the shape of the Earth today than ever before. These things have made quite a difference.

Legacy Geodetic Surveying

In measuring the Earth, accuracy unimagined until recent decades has become available from the *Global Navigation Satellite Systems, including the Global Positioning System, GNSS/GPS,* and other space-based technologies. These advancements have, among other things, reduced the application of some geodetic measurement methods of previous generations. For example, land measurement by *triangulation,* once the primary surveying technique used to extend geodetic networks of established points across vast areas across the globe, has lessened dramatically.

In the eighteenth, nineteenth, and beginning of the twentieth centuries, before the *electronic distance measurement* (EDM) devices, the measurement of long distances, now virtually instantaneous, was difficult. It was convenient then that triangulation kept the direct measurement of the sides of the triangles to a minimum. Rather the triangulation method relied heavily on the accurate measurement of the angles between the sides of large triangles, so a whole chain of *braced quadrilaterals* could be constructed (see Figure 2.1) requiring just a few measured baseline distances. When the quadrilaterals were arranged such that all vertices were intervisible, the length of each leg was verified from independently measured angles. The completed chains of quadrilaterals efficiently covered hundreds and thousands of miles, and

Detail
Idaho, Utah, Colorado, Wyoming, Nevada
and Arizona

See
Detail

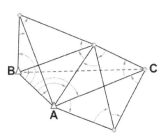

Known Data:
Length of base line AB.
Latitude and longitude of points A and B.
Azimuth of line AB.

Measured Data:
Angles to the new control points.

Computed Data:
Latitude and longitude of point C, and others
new points.
Length and azimuth of line AC.
Length and azimuth of all other lines.

FIGURE 2.1
U.S. triangulation control network map.

physical monuments were embedded in the ground at the nodes. It was through these *passive* control stations that the datum was made real. It was *realized*. The triangulation measurements also revealed the actual shape of the Earth and provided foundations on which geodesists calculated the parameters of ellipsoids, many of which are still used as references around the world today.

Ellipsoids

They each have a name, often the name of the geodesist that originally calculated and published the figure, accompanied by the year in which it was established or revised. For example, Alexander R. Clarke used the shape of the Earth he calculated from surveying measurements in France, England, South Africa, Peru, and Lapland, including M. Struve's work in Russia and Colonel Everest's in India to establish his *Clarke 1866* ellipsoid. And even though Clarke never actually visited the United States that ellipsoid became the reference for *North American Datum 1927, NAD27*. NAD27 was the foundation for geodetic surveying measurements in North America for most of the twentieth century. Despite the familiarity of Clarke's 1866 ellipsoid, it is important to specify the year when discussing it, which is true of many ellipsoids. The same British geodesist is also known for the Clarke 1858 and Clarke 1880 ellipsoids.

Supplementing this variety of regional reference ellipsoids are the new ellipsoids with wider scope, such as the *Geodetic Reference System 1980 (GRS80)*. It was adopted by the International Association of Geodesy (IAG) during the General Assembly 1979 as a reference ellipsoid appropriate for worldwide coverage. But as a practical matter such steps do not render regional ellipsoids irrelevant any more than GPS/GNSS measurements make ignorance of classical triangulation surveys prudent. Any successful GIS requires a merging of old and new data, and an understanding of legacy systems is, therefore, essential.

It is also important to remember that ellipsoidal models provide the reference for geodetic datums and contribute to the datum's definition, but they are not the datums themselves. For example, the figure for the *OSGB36* datum in Great Britain is the *Airy 1830* reference ellipsoid. The reference ellipsoid for *The European Datum 1950* is *International 1924*. The reference ellipsoid for the *German DHDN* datum is *Bessel 1841*. There are several cases where an ellipsoid was used for more than one regional datum, for example, the *GRS67* ellipsoid was the foundation for both the *Australian Geodetic Datum 1966* (now superseded by *GDA94*), and the *South American Datum 1969*. GRS80 is the reference ellipsoid for both *North American Datum 1983, NAD83*, and for the *International Terrestrial Reference Frame, ITRF*.

Ellipsoid Definition

To elaborate on the distinction between ellipsoids and datums it might help to take a look at the way geodesists have defined ellipsoids. It has always been quite easy to define the size and shape of a *biaxial* ellipsoid, an ellipsoid with two axes. It can be defined by two geometric specifications. At least it is easy after the hard work is done, that is, once there are enough actual surveying measurements available to define the shape of the Earth across a substantial part of its surface. In the past this was done by extensive triangulation surveying.

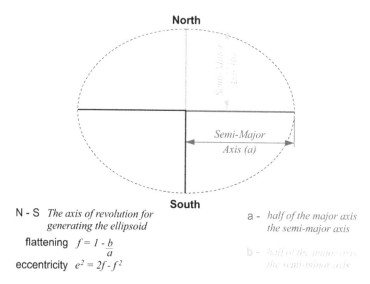

North

Semi-Major
Axis (a)

South

N - S   *The axis of revolution for*
          generating the ellipsoid

flattening $f = 1 - \dfrac{b}{a}$

eccentricity $e^2 = 2f - f^2$

a -   *half of the major axis*
      the semi-major axis

b -   *half of the minor axis*
      the semi-minor axis

FIGURE 2.2
Parameters of a biaxial ellipsoid.

The size is usually defined by stating the distance from the center to the ellipsoid's equator. This number is known as the semimajor axis, and is usually symbolized by (a) (see Figure 2.2). The shape can be described by one of the several values. One is the distance from the center of the ellipsoid to one of its poles that is known as the semiminor axis, symbolized by (b). Another parameter that can be used to describe the shape of an ellipsoid is the first eccentricity, or e. Finally a ratio called flattening, f, will also do the job of codifying the shape of a specific ellipsoid. Sometimes the reciprocal of the flattening is used instead. The definition of an ellipsoid then is accomplished with two numbers. It usually includes the semimajor and one of the others mentioned. For example, here are some pairs of constants that are usual: first, the semimajor and semiminor axes in meters; second, the semimajor axis in meters with the flattening, or its reciprocal; and third, the semimajor axis and the eccentricity.

Using the first method of specification, the semimajor and semiminor axes in meters for the Airy 1830 ellipsoid are 6,377,563.396 m, and 6,356,256.909 m, respectively. The first and larger number is the equatorial radius. The second is the polar radius. The difference between them, 21,307.05 m, is equivalent to about 13 miles, not much across an entire planet.

Ellipsoids can also be precisely defined by their semimajor axis and flattening. One way to express the relationship between the two is the formula

$$f = 1 - \frac{b}{a}$$

where:
  f is the flattening
  a is the semimajor axis
  b is the semiminor axis

The flattening for Airy 1830 is calculated as

$$f = 1 - \frac{b}{a}$$

$$f = 1 - \frac{6,356,256.910 \text{ m}}{6,377,563.396 \text{ m}}$$

$$f = \frac{1}{299.3249646}$$

The reciprocal, or inverse, flattening is then 299.3249646.

In many applications some form of eccentricity is used, rather than flattening. In a biaxial ellipsoid, the eccentricity expresses the extent to which a section containing the semimajor and semiminor axes deviates from a circle. It can be calculated as follows:

$$e^2 = 2f - f^2$$

where:
  f is the flattening
  e is the eccentricity

The eccentricity, also known as the first eccentricity, for Airy 1830 is calculated as

$$e^2 = 2f - f^2$$

$$e^2 = 2(0.0033408506) - (0.0033408506)^2$$

$$e^2 = 0.0066705397616$$

$$e = 0.0816733724$$

Figure 2.2 illustrates the plane figure of an ellipse with two axes. To generate the solid ellipsoid used to model the Earth the plane figure is rotated around the shorter axis of the two, the minor axis. The result is illustrated in Figure 2.3, where the length of the major axis is the same all along the figure's equator. This sort of ellipsoid is known as an *ellipsoid of revolution*.

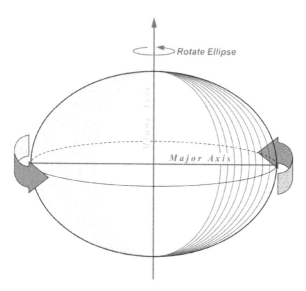

FIGURE 2.3
Biaxial ellipsoid model of the Earth.

The length of the semimajor axis is not constant in *triaxial* ellipsoids, which are also used as models for the Earth (see Figure 2.4). This idea has been around a long time. Captain A. R. Clarke wrote the following to the Royal Astronomical Society in 1860, "The earth is not exactly an ellipsoid of revolution. The equator itself is slightly elliptic."

a = 6,378,245m
b = *6,356,863.019m*
c = *6,378,033m*

Flattening
at poles $^{1}/298.3$
at equator $^{1}/30,086$

FIGURE 2.4
Krassovsky triaxial ellipsoid.

Therefore, a triaxial ellipsoid has three axes with flattening at both the poles and the equator, so that the length of the semimajor axis varies along the equator. For example, the *Krassovsky (1940)*, aka *Krasovski (1940)*, ellipsoid is used in most of the nations formerly within the USSR.

Its semimajor axis, (a), is 6,378,245 m. The flattening at the poles of 1/298.3 so axis, b, is 6,356,863.019 m and the flattening along the equator of 1/30,086 so axis c is 6,378,033 m.

Ellipsoid Orientation

Assigning two parameters to define a biaxial reference ellipsoid is not difficult. The important work of defining the orientation of the model in relation to the actual Earth is not so straightforward. The attachment of the reference model to the Earth is a step toward it becoming a geodetic datum.

The Initial Point

When it comes to fixing an ellipsoid to the Earth there are definitely two methods: the old way NAD83 and the new way. In the past, the creation of a geodetic datum included fixing the ellipsoid best suited to a region to a single *initial point* on the Earth's surface.

The initial point had to be at a known latitude and longitude, and because the coordinates of existing astronomical observatory were usually well known and long established, they were often chosen as the site of the initial point. Observatories were also convenient places to determine another prerequisite for the ellipsoid's orientation, an azimuth from the initial point to another reference point. These parameters, along with the already mentioned two dimensions of the ellipsoid itself made five in all. Five parameters were adequate to define a geodetic datum in the old way and NAD27 followed this line.

The North American Datum 1927

The New England Datum 1879 was the first geodetic datum of this type in the United States. The reference ellipsoid was Clarke 1866 mentioned earlier, with a semimajor axis, (a), of 6378.2064 km and a flattening, f, of 1/294.9786982. The initial point chosen for the New England Datum was a station known as Principio in Maryland, near the center of the region of primary concern at the time. The dimensions of the ellipsoid were defined, Principio's latitude and longitude along with the azimuth from Principio to station Turkey Point were both derived from astronomic observations, and the datum was oriented to the Earth by five parameters. Then the process of realization was begun.

Successful surveying of the first transcontinental arc of triangulation in 1899 connected it to the surveys on the Pacific coast. Other work tied in surveying near the Gulf of Mexico, and the system was much extended to the south and the west. It was officially renamed the *United States Standard Datum* in 1901.

When Canada and Mexico agreed to incorporate their control networks into the United States Standard Datum the name was changed again to *North American Datum 1913*. Further adjustments were required because of the constantly increasing number of surveying measurements, a recurring process. This growth and readjustment completed in the 1930s eventually led to the establishment of the North American Datum 1927, NAD27.

Before, during, and for some time after this period the five constants mentioned were considered sufficient to define the datum. Two were the latitude and longitude of the NAD27's initial point, Meade's Ranch, latitude 39° 13' 26.686" N and longitude 98° 32' 30.506" W. The next two parameters described the Clarke 1866 ellipsoid. The fifth and final parameter was the azimuth from the initial point to the reference point for orientation, station Waldo. It was fixed at 75° 28' 09.64".

Still other values were sometimes added to the five minimum parameters during the same era, for example, the *geoidal height* of the initial point. The assumption was sometimes made that the minor axis of the ellipsoid was parallel to the rotational axis of the Earth. The deflection of the vertical at the initial point was also sometimes considered. For the definition of NAD27, both the geoidal height and the deflection of the vertical were assumed to be zero. That meant it was often assumed that, for all practical purposes the ellipsoid and what was known as mean sea level were substantially the same,

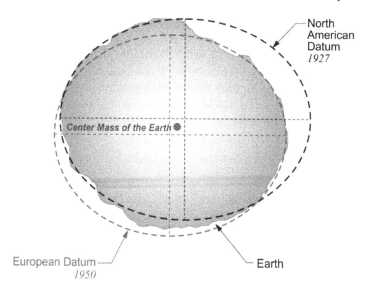

FIGURE 2.5
Regional ellipsoids.

as measurement has become more sophisticated that assumption has been abandoned.

There was one parameter that was not much considered in the creation of NAD27. It was the location of the center of NAD27's reference ellipsoid. NAD27, like most old regional datums, is not *geocentric*. Geocentric means that the center of the reference ellipsoid and the center of mass of the planet are coincident. The center of NAD27's Clarke 1866 ellipsoid was thought to reside somewhere around the center of the Earth. It worked out that they were about 236 m (~774 ft) apart as shown in Figure 2.5, but the real concerns at the time were the points on the surface of the Earth not its center. This point of view was sensible because NAD27 was a two-dimensional, horizontal datum. One consequence of this point of view is the polar and equatorial axes of older, nongeocentric ellipsoids do not coincide with the polar axis and equatorial plane of the actual Earth. The axis of the ellipsoid and the axis of the Earth were often assumed to be parallel and within a few hundred meters of each other, but not coincident.

Ground-Based Realization

Earlier there was a brief mention of the physical passive control monuments, one type of the concrete manifestation of a datum, known as its realization. Here too there is the old way and the new way. In the old way the realization of a datum required conventional measurements along the surface of the Earth from point-to-point. These ground-based methods were used to realize NAD27 and create a network of control stations set and ready to support subsequent surveying work. The physical network was necessary because the users of NAD27 could hardly have begun all their surveys from the datum's initial point in central Kansas. The *United States Coast and Geodetic Survey, USC&GS* was the forerunner of the National Geodetic Survey of today and they produced triangulation surveys that established control stations in the form of small bronze disks often set in concrete or rock. They extended the realization of NAD27 throughout the country. During that extension each measured angle and distance in the surveys had some small error. Those small errors crept into the coordinate of each new control station and built up one measurement at a time. So, as the chains of braced quadrilaterals moved forward, each rested on the errors accumulated from the hundreds and thousands of miles of surveying that preceded it. When the work was competed the results were adjusted. The adjustment distributed the accumulated errors throughout the network, but it could not and did not eliminate them.

A Systematic Change

From around the middle of the twentieth century onward, new measurement technologies initiated systematic changes to the process of realization that are still playing out. Space-based techniques that emerged in the 1960s and matured in the 1990s include: *Satellite Laser Ranging (SLR)*, *Very Long Baseline Interferometry (VLBI)*, *Lunar Laser Ranging (LLR)*, *Doppler Orbitography by Radiopositioning Integrated on Satellites (DORIS)*, *the Global Positioning System (GPS)*, *and the Global Navigation Satellite System (GNSS)*.

Today, GPS/GNSS receivers are relatively small, cheap, and easy to operate. They provide centimeter level positioning accuracy. They also eliminate the inevitability of the accumulation of error as it exists in conventional surveying such as the triangulation mentioned earlier. That systematic change makes a substantial difference in realization. Even after the advent of the first *electronic distance measurement, EDM,* devices in the 1950s had helped reveal some of the local and regional distortions in the NAD27 conventional surveying still involved the determination of horizontal coordinates by measuring from point-to-point along the Earth's surface. These methods cause errors to accumulate because the error in each point and line in the network must be built on all the errors in all of the previous points and lines. Space-based measurement systems need not proceed that way. Accumulation of error is not inevitable because the measurement of each point in the network can originate from the same source. Each point so determined certainly has error, but that error need not be added to subsequent points.

Consider two points 30 miles apart whose positions are established using the same constellation of GPS/GNSS satellites. In that case the accuracy of one station's position and the accuracy of the other station's position 30 miles away do not depend on each other because they were both measured directly from the same source. However, if the positions of those two stations were established by a thirty mile long triangulation or traverse survey the situation would be different. The accuracy of the second station would not only depend on the accuracy of the first, it would also depend on the accuracy of every single measurement made in the connecting survey between them.

The 1980s were early days in the use of GPS, and NAD27 was the datum available. In those days, surveyors would begin their GPS surveys and calculations from an established fixed official passive network control station with published NAD27 coordinates. They would then move on to set and position a new project point, or points, that they required. Once that was done, they would check their work by tying in another known official passive network control station that also had an NAD27 published coordinate. This checking in would frequently reveal that their GPS work had created coordinates that did not match the published NAD27 coordinates at the official known stations. When that happened even though the surveyor's GPS measurements were more accurate than the published coordinates that had been established by ground-based surveying long ago, it was not feasible to alter the published coordinates of the national network. The surveyors had no choice but to adjust the more accurate GPS surveyed measurements to match the less accurate NAD27 published network control coordinates. Adjusting satellite-derived measurements to fit the less accurate coordinates available from NAD27 was untenable and revealed that the old datum could not support the improved measurement technology. A new fundamentally different datum was needed.

The North American Datum of 1983

In 1986 the North American Datum of 1983, NAD83, replaced the North American Datum of 1927. It took more than 10 years to readjust and redefine the horizontal coordinate system of North America into NAD83. More than 1.7 million weighted surveying observations were involved, some 30,000 EDM-measured baselines, 5,000 astronomic azimuths; about 655 Doppler stations were positioned using the *TRANSIT* satellite system and about 112 Very Long Baseline Interferometry (VLBI) vectors. The North American Datum of 1983, NAD83 can be said to be the first civilian coordinate system established with some space-based measurements. However, little if any, GPS data were used in the original establishment of NAD83. Still GPS/GNSS and other satellites orbit with the center of mass of the Earth at one focus of the elliptical paths they follow. So, an attempt was made to make the GRS80 reference ellipsoid for the new NAD83 geocentric, but the systems used were inadequate to rigorously determine the center of mass of the Earth.

When NAD83 coordinates were implemented across the United States, coordinates shifted. The smallest differences occur in the middle of the United States and as long as the area involved is small the shift is relatively constant. However, as the area considered grows, there are significant but variable shifts between NAD27 coordinates and NAD83 coordinates. The differences can grow to approximately +1.5 s in latitude, 150 ft north–south, but the change is generally larger east–west up to +5.0 s in longitude, 400 ft east–west. The longitudinal shifts are actually a bit larger than that in Alaska, ranging up to 12.0 s in longitude.

If the reference ellipsoid for both NAD27 and NAD83 had been the same, the changes in the coordinates would not have been that large. For example, had NAD83 coordinates had been projected onto the same Clarke 1866 ellipsoid oriented as it had been used for NAD27 the change in coordinates would have been smaller. If there had not been such a systematic change in measurement technology the shifts in the coordinates may have been more consistent.

The evolution of NAD83 has continued since it was introduced in 1986. The original version of the datum is often written as NAD83 (1986). However, in 1986 GPS was not fully operational. That would happen 11 years later. So, it took some years for it to become clear that there were local and regional distortions in the realization of the new datum too. In other words, there were significant discrepancies between GPS/GNSS positions and the NAD83 (1986) published coordinates. That problem and a few others needed a solution.

The NAD83 (1986) stations, the million or more passive marks set across the United States, could be occupied by conventional surveying instruments, but they were often not suitable for GPS/GNSS observations because of overhead obstructions, they could be difficult to reach and they were irregularly spaced across the network.

In an effort to alleviate some of these problems, from 1989 to 1997 state-by-state cooperative surveying was done. The work was shared by the states and the *National Geodetic Survey, NGS*, an office of *NOAA's National Ocean Service*. The first objective was to improve the accuracy of the geographic coordinates on the passive control. As vertical measurement technology had also improved a second round of work was done to upgrade the ellipsoid heights too. The result was called *NAD83 High-Accuracy Reference Network (HARN)*. It was an improvement. The stations were more regularly spaced, always within 20–100 km, 12–62 miles, of one another. They were more accessible. That work provided many more GPS measurements of high accuracy such as A-Order (5 mm + 1:10,000,000) and B-Order (8 mm + 1:1,000,000). It was at this stage that NAD83 moved from being a two-dimensional horizontal datum as was NAD27 to a three-dimensional, latitude, longitude, and ellipsoidal height, geometric datum.

NAD83 has been upgraded many times since. For example, the inauguration of the system of *Continuously Operating Reference Stations (CORS)* resulted in an improved realization known as *NAD83 (CORS96)*. They are now augmented by *active* stations, known as CORS. Active stations are similar to the passive marks, at fixed, official known locations with published coordinates, but they differ in a way that GPS/GNSS receivers are permanently affixed to each one. The receivers on CORS are continuously collecting satellite-positioning data and that data are made available to the public over the Internet. GPS/GNSS signals collected and archived at these sites of known position provide base data that serve as the foundation for the correction and improvement of all sorts of positioning, navigation, and timing work in the United States and its territories. Today, when GPS/GNSS data are collected at an unknown station it can be processed with data from a CORS station to produce positions that have a few centimeter-level accuracy in relation to the *NSRS*. NSRS stands for *National Spatial Reference System*.

NAD83 (CORS96) was followed by subsequent adjustments such as *NAD83 (NSRS 2007)*. This adjustment included the GPS observations on passive stations and was constrained to the CORS network. The goal was the removal of the state-to-state coordinate inaccuracies that developed in the HARN work, but it could only address the 70,000 passive marks included in the adjustment. The most recent realization is the result of a national adjustment and is known as *NAD83 (2011)*.

Space-Based Realization

Modern realizations, the positioning of new points on the surface of the Earth, are accomplished with space-based measurements that incorporate orbiting satellites, quasars, and reflectors on the Moon. This remarkable change is accompanied by a new lexicon that is used to describe the realizations they produce. Today, the abstract parameters on which a system to measure positions on the Earth is built is known as a *Terrestrial Reference System, TRS.* It is a set of constants with which a system can be abstractly defined. The realization of that Terrestrial Reference System is called a *Terrestrial Reference Frame, TRF.*

The Terrestrial Reference Frame

As it is an abstract, a TRS is errorless but a TRF is certainly not. A TRF is built from coordinates derived from actual measurements. Actual measurements, even by means of space-based systems, always contain errors. Therefore, the coordinates that make up a TRF contain errors too, though increasingly small ones. A Terrestrial Reference Frame includes positions on the Earth's surface with known coordinates. They allow users to do real work in the real world, so it is important that they are easily accessible and their values published or otherwise easily known.

Comparing Terrestrial Reference Systems and Terrestrial Reference Frames

Instead of speaking of Terrestrial Reference Frames (TRFs) as separate and distinct from the Terrestrial Reference Systems (TRSs) on which they rely, the word datum is still often used to describe both. Avoiding this could prevent a good deal of misunderstanding.

The relationship between two TRSs can be defined without ambiguity by comparing the exact parameters of each. For example, comparing the respective semimajor axes and flattening of two biaxial ellipsoids, the difference between them is as clear and concise as the numbers themselves. It is easy to express such differences in absolute terms. However, such straightforward comparison is seldom the important question in day-to-day work.

On the other hand transforming coordinates from one TRF to another is an almost daily concern.

The relationship between the TRSs is easily defined; the relationship between the TRFs is much more problematic in part because a TRF cannot be a perfect manifestation of the TRS on which it is based.

World Geodetic System 1984

There is an important global geocentric TRF. The *World Geodetic System 1984 (WGS84)* is the geodetic reference system used by GPS. When WGS84 was originally rolled out in 1987 it had been developed for the *United States Defense Mapping Agency, DMA.* The agencies name has changed over the years. It became the *National Imagery and Mapping Agency, NIMA* and today it is known as the *National Geospatial-Intelligence Agency, NGA.* GPS receivers compute and store coordinates in terms of WGS84. However, the microcomputer within receivers can and often do transform them to other reference systems defined by the user.

TRANSIT

WGS84 was initially realized with observations from the *TRANSIT* satellite system. TRANSIT whose positions had accuracies in the 1 to 2 m range was an Earth-orbiting system that combined satellite technology and the Doppler Effect. The idea originated in 1957 when experimenters at Johns Hopkins University's Applied Physics Laboratory, Laurel, Maryland found that the Doppler shift of Sputnik's signal provided enough information to determine the exact moment of its closest approach to the earth. This discovery led to the creation of the *Navy Navigational Satellite System, NNSS* or *NAVSAT*, and the subsequent launch of six satellites specifically designed to be used for navigation of military aircraft and ships. The system had its first successful test in 1960. This system was operational for the military in 1964 but classified. It was declassified in 1967 and was widely used in civilian hydrographic and geodetic surveying for many years until it was switched off on December 31, 1996. Subsequent WGS84 realizations have been based on GPS data and have steadily improved. Those improvements have been tagged with *GPS weeks*.

Global Positioning System Weeks

GPS weeks and seconds of a week are counted from the midnight between January 5, 1980, which was a Saturday, and January 6, 1980, a Sunday. That was the beginning of week zero. Each new GPS week begins at the Saturday/Sunday transition and each new realization of WGS84 derived from GPS data since is tagged with a parenthetical GPS week. The first enhancement of WGS84 was in 1994 on GPS week 730 and was named WGS84 (G730). The next was a couple of years later in 1997. It became known as WGS84 (G873). Then there was WGS84 (G1150) in 2002 and WGS84 (G1674) in 2012. Because the week of October 13, 2013 was GPS week 1762 the latest realization rolled out by the United States Department of Defense on October 16, 2013 is WGS84 (G1762).

WGS84 as originally rolled out in 1987 was nearly identical with NAD83 (1986) and even though WGS84's ellipsoid and the GRS80 ellipsoid are very similar. The subsequent realizations of WGS84 have caused it to drift farther and farther from NAD83. Today, the difference between a position in NAD83 (2011) and a position in WGS84 (G1762) can approach 1 or 2 m, ~3 ft to ~6 ft. As it moved away from NAD83, WGS84 has been kept virtually coincident with the International Terrestrial Reference Frame, ITRF. In order to explain a bit about ITRF it is necessary to start with the *International Earth Rotation and Reference Systems Service, IERS*.

The International Earth Rotation and Reference Systems Service

The IERS is an organization that began operations in Paris at the beginning of 1988 under the auspices of the International Astronomical Union, IAU, and the International Union of Geodesy and Geophysics, IUGG. The IERS defines the International Terrestrial Reference Frame. As part of that process the IERS has formally defined the components of a three-dimensional Cartesian coordinate system. Every position on the planet, and in its vicinity, can have a unique coordinate in this system. Positions derived directly from GPS/GNSS observations are expressed in the x, y, and z coordinates of this system. It is known as the *Conventional Terrestrial Reference System, CTRS,* or just the *Conventional Terrestrial System, CTS* (Figure 2.6).

The Conventional Terrestrial Reference System

The origin is the center of mass of the whole Earth including oceans and atmosphere, the geocenter. However, the relationship between the surface of the Earth, its center, and even its axis of rotation is not constant and unchanging. The Earth wobbles in motions known as *precession* and *nutation*. Precession is the long-term movement of the polar axis. It moves in a circle with a period of something approximating 25,800 years. At the moment the planet's spin axis almost points to Polaris, but in 14,000 years or so it might

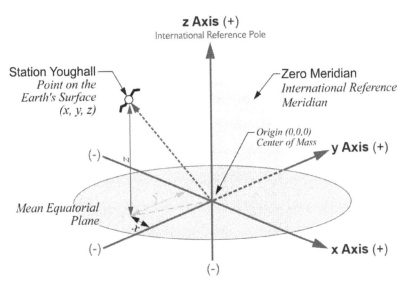

FIGURE 2.6
A three-dimensional Cartesian coordinate in the CTRS.

point to Vega. Nutation is the movement of the Earth with a cycle of about 18.6 years, mostly attributable to the moon. The Earth's rotation rate also varies. It is a bit faster in January and slower in July and then there is the wandering of the Earth's axis of rotation relative to the Earth's surface, called *polar motion.*

International Earth Rotation and Reference Systems Service Reference Pole

Polar motion is a consequence of the actual movement of the Earth's spin axis as it describes an irregular circle with respect to the Earth's surface. The circle described by this free Eulerian motion of the pole has a period of about 435 days or so. It takes approximately that long for the pole to complete a circle that has a diameter of about 12 m to 15 m, ~39 ft to 49 ft. This part of the polar motion is known as the *Chandler period,* named after American Astronomer Seth C. Chandler who described it in papers in the *Astronomical Journal* in 1891. The motion of the Earth's spin axis from 2000 to 2016 is shown generally in Figure 2.7.

Therefore, one can say that the Earth has a particular axis of rotation for an instant before it changes in the next instant. Within all this motion how do you define a stable, static pole? One way is to choose a moment in time and fix it at that instant. That was how it was done. A moment was chosen by the *Bureau International de l' Heure, BIH* and also adopted by the *International Earth Rotation Service, IERS.* It was the moment of midnight on the transition between New Year's Eve 1983 and January 1, 1984 (UTC). That moment

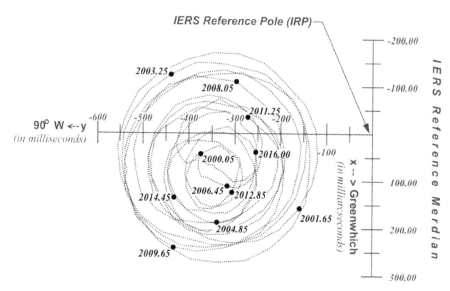

FIGURE 2.7
Motion of the Earth's spin axis 2000–2016.

is also known as an *epoch* and can be written 1984.0. That internationally agreed upon position of the pole is known as the *IERS Reference Pole, IRP.*

The x-axis is a line from that geocenter through its intersection at the *IERS Reference Meridian (IRM)*, also known as the zero meridian, with the internationally defined conventional equator. The y-axis is extended from the geocenter along a line perpendicular from the x-axis in the same mean equatorial plane. That means that the positive end of the y-axis intersects the actual Earth in the Indian Ocean. In any case, they both rotate with the Earth around the z-axis which is a line from the geocenter through the internationally defined pole known as the IERS Reference Pole (IRP). The three dimensional Cartesian coordinates (x, y, z) derived from this system are sometimes known as *Earth-Centered Earth-Fixed (ECEF)* coordinates.

In this three-dimensional coordinate system the x-coordinate is a perpendicular distance from the plane of the y-axis. It is always positive from 90° W longitude to 90° E longitude. In the remaining 180° of longitude the x-coordinate is negative.

The y-coordinate is a perpendicular distance from the plane of the x-axis. It is always positive in the Eastern Hemisphere and negative in the Western Hemisphere.

The z-coordinate is a perpendicular distance from the plane of the equator. It is always positive in the Northern Hemisphere and negative in the Southern Hemisphere.

Here is an example, the position of the station Youghall expressed in three-dimensional Cartesian coordinates of this type expressed in meters, the native unit of the system:

$$x = -1564831.1855$$

$$y = -4605604.7477$$

$$z = 4115817.6900$$

The x-coordinate for Youghall is negative because its longitude is west of 90° W longitude. The y-coordinate is negative because the point is in the Western Hemisphere. The z-coordinate is positive because the point is in the Northern Hemisphere.

This ECEF coordinate system is *right handed* and orthogonal. It can be illustrated as shown in Figure 2.8; the horizontally extended forefinger of the right hand symbolizes the positive direction of the x-axis. The middle finger of the same hand extended at right angles to the forefinger symbolizes the positive direction of the y-axis. The extended thumb of the right hand, perpendicular to both of them, symbolizes the positive direction of the z-axis.

The axes are stable, clearly defined, and rotate with the Earth, but what about the planet's surface? It is well known that the tectonic plates and the stations on them move with respect to the geocenter and the planet's interior. The absolute plate motion is virtually eliminated by a *No-Net-Rotation; NNR*

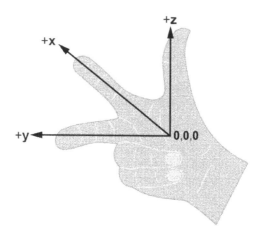

FIGURE 2.8
The ECEF coordinate system is right handed and orthogonal.

condition is imposed onto the system by the IERS. NNR employs a weighted average of all of the world's plate velocities within which the motions of the surface plates with respect to the geocenter and the planet's interior are represented. That means that the total angular momentum of all tectonic plates together should be reduced to zero.

It works well but the NNR condition does not eliminate relative plate velocities. The relative motion between the approximately 20 large tectonic plates that make up that surface is not handled by NNR. Stations on separate plates can move as much as 150 mm per year simply because the ground on which one station stands is slowly shifting in relation to the ground at another. Hence changes in the three-dimensional Cartesian coordinates are inevitable. It is this fact, among others, that led the IERS introduced velocity to the coordinates.

The International Terrestrial Reference Frame Realizations

IERS's first realization of the ITRS was the *International Terrestrial Reference Frame of 1988* (*ITRF88*). Nearly every year since, IERS has published a list of new and revised ITRF positions with their velocities around the world. The IERS released ITRF89, ITRF90, ITRF91, ITRF92, ITRF93, ITRF94, ITRF95, ITRF96, and ITRF97. Please note that the digits appended to ITRF represent the year up to which the data sets have been used in the realization.

After the 1990s ITRF2000, ITRF2005, ITRF2008, and ITRF2014 followed. The period between realizations lengthened due to the lessening in the size of the positional updates. ITRF93 differs from ITRF00 by approximately 3 cm (0.1 ft) at the surface of the Earth. However, the transformation parameters between ITRF2005 and ITRF2000 are smaller. The difference in the x-coordinate is approximately 0.1 mm (0.0003 ft); the difference in the y-coordinate

TABLE 2.1

Reference Frame Equivalencies

Introduced	Realization	Datum Epoch	For All Practical Purposes Equivalent To:	For All Practical Purposes Equivalent To:
1987	WGS84 (1987)	1984	NAD83(1986)	
June 29, 1994	WGS84 (G730)	1994.0	ITRF91	
January 29, 1997	WGS84 (G873)	1997.0	ITRF94	
January 20, 2002	WGS84 (G1150)	2001.0	ITRF00	
February 8, 2012	WGS84 (G1674)	2005.0	ITRF08	IGS08
October 16, 2013	WGS84 (G1762)	2005.0	ITRF08 and ITRF2014	IGS08

is approximately 0.8 mm (0.003 ft) and the difference in the z-coordinate is approximately 5.8 mm (0.02 ft).

In all cases the ITRF realizations were based on a weighted combination of stations positions and *Earth Orientation Parameters* (*EOPs*) provided by space-based geodetic measurements from SLR, VLBI, and so on taken at sites all around the world. The number and evenness of distribution of ITRF sites and collocations has increased over the years. Collocations are sites where two or more instruments are in close proximity. ITRF88 was derived from observations at about 100 sites and 22 collocations and for the ITRF00 realization the number of sites increased to about 800 stations at 500 sites and 101 collocations.

The ITRF stations are moving, and recognized to be doing so. Therefore, they each are described by a position and a velocity and every position, every set of coordinates in these realizations refer to the stations position at a particular moment. This moment is known as the *Reference Epoch* (*RE*). For example, the RE for ITRF05 is 2000.0, or more specifically January 1, 2000 at exactly 0:00 UTC as shown in Table 2.1. Even stations on the most stable part of the North American plate are in horizontal motion continuously at rates that range from 9 to 21 mm every year. Although it must be said that these changes in coordinates are very slight indeed, they are changes nonetheless and ITRF accounts for them.

The North American Datum of 1983, World Geodetic System 1984, and International Terrestrial Reference Frame

As mentioned earlier as it moved away from NAD83, WGS84 has been kept virtually coincident with the International Terrestrial Reference Frame, ITRF. The situation is illustrated in Figure 2.9.

The expression of a single position in North Dakota, Medora 5, in several reference frames can cause this CORS to appear to be in several places at once. The two realizations of NAD83 shown in Figure 2.9 are 0.08 ft apart.

FIGURE 2.9
Station Medora 5 in several reference frames.

The Medora 5 position in ITRF00 is about 0.38 ft from its ITRF08, IGS08, and WGS84 position. However, the distance between the WGS84, ITRF positions and the NAD83 positions is more than 1 m, approximately 4 ft.

WGS84 and ITRF have grown together, whereas WGS84 and NAD83 have grown apart because WGS84 realizations using space-based data, including GPS/GNSS, has brought the center of the WGS84 reference ellipsoid very close indeed to the actual center of mass of the Earth. Please recall that the geocenter is one focus of the satellites orbits and the origin of the measurements derived from them. It is a good idea for the reference frame to be centered there too.

For all practical purposes the following are equivalent, at the 10 cm level: WGS84 (G730) and ITRF91, WGS84 (G873) and ITRF94, WGS84 (G1150) and ITRF00, and WGS84 (G1674) and ITRF08. In fact, WGS84 (G1762) is likely to agree within a centimeter or so with ITRF2014 and ITRF08.

Please note the presence of the reference frame known as *International GNSS Service 08, IGS08* in Table 2.1.

International Global Navigation Satellite Systems Service 08

NGS began utilizing ITRF data in March of 2002 and upgraded its published NAD83 positions and velocities of the CORS stations to be equal to the ITRF2000 positions and velocities. However, on September 6, 2011 NGS replaced the ITRF2000 realization with IGS08 epoch 2005.0. IGS08 is GPS/GNSS determined and is virtually equivalent to ITRF08.

Here is an illustration of the application of IGS08 epoch 2005.0 data. The coordinate values for the position of station Alternate Master (AMC2) in Colorado Springs, Colorado are shown here as retrieved from the NGS database. Here is the IGS08 position with its velocities in each dimension:

$$x = -1248596.192 \text{ m} \quad Vx = -0.0146 \text{ m/yr}$$

$$y = -4819428.211 \text{ m} \quad Vy = 0.0009 \text{ m/yr}$$

$$z = 3976505.997 \text{ m} \quad Vz = -0.0049 \text{ m/yr}$$

Here is the NAD83 (2011) epoch 2010.0 position of the station in three-dimensional Cartesian coordinates with velocities in each dimension.

$$x = -1248595.509 \text{ m} \quad Vx = 0.0021 \text{ m/yr}$$

$$y = -4819429.537 \text{ m} \quad Vy = 0.0018 \text{ m/yr}$$

$$z = 3976506.059 \text{ m} \quad Vz = -0.0005 \text{ m/yr}$$

The question, *Where is station AMC2?* might be more correctly asked as, *Where is station AMC2 now?* The latter question can be answered by calculating new positions for the station based on its velocities. To do that IGS08 ECEF coordinates can be transformed into geographical coordinates.

Three-dimensional Cartesian coordinates and geographical coordinates with ellipsoidal heights can be calculated from one another. Here are the expressions for deriving the three-dimensional Cartesian coordinates from latitude, longitude, and height above the ellipsoid:

$$x = (N + h) \cos \varphi \cos \lambda$$

$$y = (N + h) \cos \varphi \sin \lambda$$

$$z = \left(\frac{Nb^2}{a^2 + h} \right) \sin \varphi$$

where:
  φ is the latitude
  λ is the longitude
  h is the height above the ellipsoid
  N is the east–west local radius of the reference ellipsoid in meters

$$N = a^2(a^2 \cos^2 \varphi + b^2 \sin^2 \varphi)^{-\frac{1}{2}}$$

where:
  a is the semimajor axis
  b is the semiminor axis

There is no reference ellipsoid specified for IGS08. Nevertheless, the GRS80 ellipsoid is recommended. This inclusion along with velocity introduces four-dimensional geographical coordinates: latitude, longitude, ellipsoidal height, and velocity.

In this way the location of the AMC2 station can also be stated in latitude, longitude, and height above the ellipsoid and velocity. Here is the AMC2 IGS08 position:

Latitude = 38° 48' 11.24802" N   northward = –0.0056 m/yr

Longitude = 104° 31' 28.53765" W   eastward = –0.0144 m/yr

Ellipsoid height = 1911.395 m   upward = –0.0009 m/yr

Here is the AMC2 NAD83 (2011) position:

Latitude = 38° 48' 11.22697" N   northward = 0.0010 m/yr

Longitude = 104° 31' 28.49648" W   eastward = 0.0016 m/yr

Ellipsoid height = 1912.300 m   upward = –0.0021 m/yr

Predictably the distance between the NAD83 (2011) coordinate for AMC2 and the IGS08 (ITRF08) coordinate is 1.1867 m, 3.89 ft. The ellipsoidal height difference between the two is 2.97 ft, 0.905 m.

Replacement of the North American Datum of 1983

NGS plans to replace NAD83 with a new semidynamic terrestrial reference frame for North America. The implementation of the plan is currently scheduled for 2022. The interim period is intended to smooth the transition to the new paradigm and completion of measurements critical to creation of the new geopotential datum. The name of the portion of the new system is not yet certain but the portion that will be fixed to the North American tectonic plate may be called the *North American Terrestrial Reference Frame of 2022 (NATRF2022)*. One objective of the plan is the rectification of some long running discrepancies between the orientation of the TRF in North America and the space-based systems now used to realize it.

The orbits of GPS/GNSS satellite constellations are geocentric. The global reference frames ITRF (2008), IGS08, and WGS84 (G1762) are also geocentric. Since GPS/GNSS is the dominant surveying utility worldwide the equivalent orientation of these reference frames makes sense. However, NAD83 is not geocentric. In fact, a comparison of NAD83 and ITRF88 three-dimensional coordinates revealed that the difference between their centers was more than 2.2 m (~7 ft). Nevertheless, all realizations of NAD83 have been constructed on this same nongeocentric foundation. It follows, that NAD83 (2011) positions diverge systematically from any positions that rest on the geocentric reference frames. This positional difference is especially troublesome as the accuracy of so-called mapping grade GPS work done in WGS84 (G1762), a geocentric TRF, becomes more and more accurate. Soon inexpensive GPS/GNSS multiconstellation receivers will be capable of sub-meter accuracy without differential correction. The discrepancy of up to, and perhaps more than, 2 m (6.5 ft) between NAD83 (2011) and WGS84 (G1762)

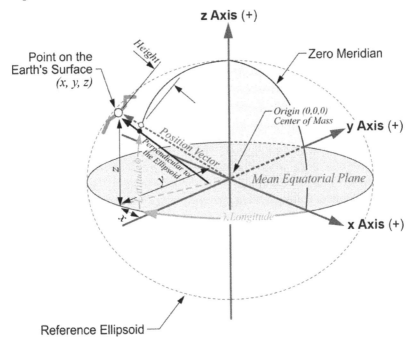

FIGURE 2.10
ECEF and geographic coordinates.

FIGURE 2.11
Horizontal shift from NAD832011(2010.00) to ITRF2008(2020.00) in meters.

will be troublesome and further exacerbated by some software products that still insist that NAD83 and WGS84 are equivalent, which has not been true since 1987 (Figure 2.10).

The horizontal shift in meters from NAD83(2011) epoch 2010.00 to ITRF (2008) 2020.00 illustrated in Figure 2.11 is probably representative of the shift that will occur in 2022 when the new TRF is implemented. That shift varies from approximately 0.7 m (2.3 ft) to approximately 2.6 m (8.5 ft) in the Continental United States, the largest shift being along the westernmost portion of California.

There will also be a change in the public's access to the new TRF. When implemented the new frame's coordinates will primarily exist on active control rather than on passive control.

In the past official the static coordinates published by the NGS were assumed to be unchanging. The measurement systems available at the time created sufficiently reliable positions on passive monuments. Their constant slow movement was not apparent then because the errors in the measurement systems exceeded the extent of the motion. That is no longer the case.

Measurement technology has improved dramatically. It now reveals the persistent movement. This fact has diminished the reliability of the passive control monuments and makes it hard to justify the extraordinary resources that would be necessary to keep their coordinates up-to-date. As the current passive control network becomes unsustainable, in some instances, the new TRF may be directly available to users primarily via active control.

The new paradigm will also alter the roles and responsibilities that have long supported the NSRS. For many years the NGS has established and maintained the passive control network under the new plan that will end. That responsibility shifts onto the users. If users need passive control monuments they will need set, and tie them to the NSRS themselves, but such control will not be considered an official part of the NSRS.

NGS will continue to provide users with utilities to process their data, that is, *Online Positioning User Service (OPUS)*, which will be augmented with new services including tools to transform their coordinates from NAD83 to the new TRF and tools to assign velocities to them. Currently NGS does not report motion on passive control. Users will know that when they have a velocity on passive marks they are working in the new reference frame and not in NAD83. When users know the movement of the monument in time they will have the option to adopt a specific epoch for their work if they wish. In other words, a user may choose to work in a nondynamic system or, using the active control, work in the fully dynamic system.

These changes, that is, a truly geocentric reference frame and the restriction of access to the reference frame through active control, will allow the NGS to reach their goals. Those goals include publishing CORS coordinates daily; continuous tracking changes in the active control network and supporting development of real-time networks.

The NGS is also working on a new vertical geopotential datum that will replace the *North American Vertical Datum 1988, NAVD88*.

Transforming Coordinates

Transformations are mathematical mechanisms used to move coordinates from one reference frame to another, and there are several methods. Transformations from realizations of one old style classical datum to another can be accomplished with the same mathematics used to transform them from one modern TRF to another. However, the parameters used in a classical datum transformation are likely to be much larger than those used to transform from one TRF to another TRF.

Transformations versus Coordinate Conversions

Transformations ought to be distinguished from *coordinate conversions*. Coordinate conversion is usually understood to mean the reexpression of coordinates from one format to another while both rest on the same TRF or the same classical datum. For example, the conversion of plane UTM WGS84 (G1762) coordinates to geographic coordinates, still in WGS84 (G1762), would be a coordinate conversion not a transformation.

Transformation usually means coordinates move from an original TRF to a different target TRF. Changing a positions expression from a geographic coordinate in NAD83 (2011) to a geographic coordinate in WGS84 (G1762) would be a transformation. Complications to that transformation would be the differences in the flattening between the ellipsoids of reference for NAD83 (2011), GRS80, and the ellipsoid of reference for WGS84 (G1762), and the differences in the location of their centers. Other transformation difficulties can include the orientation of the axes of the original and target TRFs. They may require rotation to coincide. The scale of the distances between points on one TRF may not be at the same scale as the distances on the target TRF and so on.

Integrity

A proper transformation will not degrade the accuracy of the coordinates themselves. However, if the distance between point A and point B is incorrect in the original system it will be just as wrong when it is transformed into a target system, no matter how sophisticated the mathematics. The accuracy with which the coordinates represent the relative positions of the actual points on the Earth cannot be improved by conversion or transformation.

Transformations work best if the points involved have been measured in both the original TRF and the target TRF. These are usually called common points. A common point has, of course, one coordinate in the original TRF and an entirely different coordinate in the target TRF; still both represent the same position on the Earth. The more common points that are available in a transformation the better. The accuracy and the distribution of these coordinates in both the original and the target TRF is also an important factor in the integrity of the outcome. Accuracy varies, particularly in the realizations of old classical datums. For example, triangulation accuracy improved as instrumentation, field procedures, and adjustment techniques evolved. It follows that transformations between two old classical datums that were established by triangulation is best defined by using points that are part of the original main triangulation arcs and common to both systems. Such care in choosing the points used in a transformation are particularly important when a large area is involved. The point selection affects the results no less than the actual calculation of the transformation.

While it is true that transformation work is typically left to computer applications today, it is, at best, unwise to accept the results uncritically. Checking is necessary. For example, some of the surveyed common points in the target TRF can be excluded from the points used and then be compared to the transformed coordinates after the transformation is completed to evaluate the consistency of the operation after the fact.

What follows is an outline of some of the qualities of some typical transformation methods.

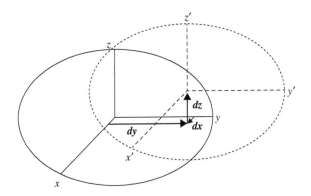

FIGURE 2.12
Molodenski transformation.

Molodenski Transformation

This method is named after the mid-twentieth century Russian physicist, M.S. Molodenskij. The Molodenski transformation is sometimes known as the 3-parameter or 5-parameter transformation. It is used on-board some GPS/GNSS receivers. In fact, the growth in the utilization of GPS/GNSS has probably increased the number of computer applications relying on this method. In this context, it is often implemented to transform coordinates from WGS84 into a local projection (Figure 2.12).

The Molodenski transformation is simple in conception and available in many standard GIS software platforms as well. It rests on shifts to the three geocentric coordinates that are applied directly to geographical coordinates. It usually requires ellipsoidal parameters from the original and the target system as well and the size of the shift in the x, y, and z geocentric coordinates. In other words, it uses simple straight-forward formulas to shift the origin from the original TRF to the target TRF along the x, y, and z axes, (Δx, Δy, Δz) based on the averaged differences between the x, y, and z coordinates of the previously mentioned common points. There is no scaling or rotation in this method.

Like all such mathematical operations the worth of this sort of transformation is dependent on the consistency of the coordinate values available, but at best it can only produce a transformation with moderate accuracy. The size of the area being transformed bears on the accuracy of the transformation.

The Molodenski transformation is based on the assumption that the axes of the original ellipsoid and the target ellipsoid are parallel to each other. That is seldom true, but if the work involved is across a small area the effect of the assumption may be insignificant. However, as the size of the area grows, so does the inaccuracy of this method of transformation. In short, the Molodenski method is satisfactory if some of the work requires modest accuracy, but rotation and scale parameters are needed for more precise work (Figure 2.13).

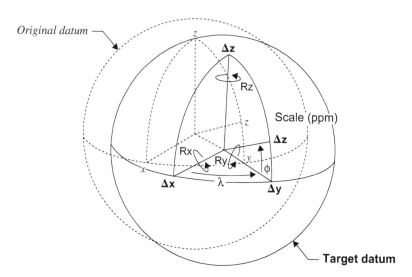

FIGURE 2.13
Translation and rotation.

Seven-Parameter Transformation

This method is also known as Helmert or Bursa–Wolf transformation. It bears remembering that TRF transformations do not improve the accuracy of the coordinates they transform. They cannot do that. However, when the number of parameters considered is increased the result is an improvement in the fit of the coordinates in the target TRF.

To transform from one geocentric TRF to another, one could use the seven parameters of the Bursa–Wolf approach; 3 translations, 3 rotations, and 1 scale factor. The sum of the x, y, and z translations accomplish the shifting of the origin so that the origins of the original TRF and the target TRF match. The shifts are usually expressed as Δx, Δy, Δz or Dx, Dy, and Dz. But they are sometimes shown as u, v, and w. In any case, the three shift distances are specified in meters. Their objective is to shift the ellipsoid along each of its three axes.

Then through the rotation of each of the axes the original TRF and the target TRF axes are made parallel to one another. The three rotation parameters of the x, y, and z axes are symbolized by Ex, Ey, and Ez or rx, ry, and rz. Though they are sometimes shown as ε_x, ε_y, ε_z or ε, ψ, and ω. The three rotation parameters specify the angles. The angles are usually less than 5 arc-seconds and are calculated by producing a combined rotation matrix. Finally, the transformation is scaled. The scale factor is usually calculated in parts per million.

This method is also known as the three-dimensional Helmert, three-dimensional conformal, or three-dimensional similarity. The seven-parameter transformation ought to start with at least three coordinates of points that are common to the original and the target system; more are better.

Again the quality of the results are dependent on the consistency of the set of common coordinated points utilized by the original and target side of the work. This transformation is available in many GIS software packages and its accuracy is better than that available with the Molodenski transformation. The seven-parameter transformation does require heights for the common coordinated points in both the original and the target systems, and the results depend on the consistency of the coordinates in both systems.

Surface Fitting

The previous transformations were based on equations. Surface fitting, illustrated in Figure 2.14, is based on a transformation grid. It is also known as bilinear gridded interpolation, or the grid-based interpolation method.

This is the best approach to TRF transformation. The United States, Canada, and Australia use the method. In 1986 when the TRF changed in the United States it was clear there were no constant values that could easily move the geographical coordinates from the original, NAD27 on the Clarke 1866 ellipsoid, to the target TRF, NAD83 on the GRS80 ellipsoid. Among the reasons were the centuries of conventional surveying that introduced unavoidable inconsistencies into the original coordinates. Under the circumstances it would have been unworkable to transform the NAD27 coordinates into NAD83 with the Molodenski method or the seven-parameter method. What was needed was an approach that was more fine-tuned and specific.

The *North American Datum Conversion Utility* (*NADCON*) was developed. The program uses grids and an error averaging strategy based on real data. When a coordinate is input for transformation the necessary shift calculations are based on a grid expressed in a database, which contains the shifts at tens of thousands of points in an extensive grid network. This grid of control points has shifts that are known, and that information is used to estimate the shift at other locations.

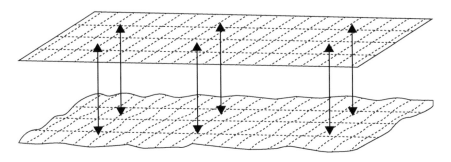

FIGURE 2.14
Surface fitting.

This approach is based on the idea that it is possible to find the necessary shift in a coordinate to transform it from its original TRF to a target TRF by interpolation from known shifts for a number of control points in the same area. For example, a particular geographic coordinate that is to be transformed will fall within a grid cell that has points of known shifts at each of its four corners. The application of a bilinear interpolation algorithm can thereby derive the necessary shift at the given point. The interpolation uses two grid files: one for the shifts in latitude and one for the shifts in longitude. This rubber-sheeting method is good, but it requires an extensive grid database to be successful. Building such a database needs the devotion of significant resources and almost certainly the auspices of governmental agencies. It is a minimum curvature method and is probably the most popular transformation routine in the United States. It is a surface fitting type of transformation.

The software behind the surface-fitting transformation is not based on simple formulas, but it can be operated from a simple user interface that emphasizes simple shifts in latitude and longitude. Several points common to the original and the target coordinate systems are used in the surface-fitting method.

On the positive side the surface fitting transformation is quite accurate and often driven by a simple user interface. It is integral to many standard GIS software packages. On the other hand it is mathematically complex and requires that the original and the target coordinate systems have many common points.

The number of common points available and the accuracy required are important considerations in choosing the appropriate transformation method. The surface fitting or grid-shift techniques like that used in NADCON provide the best results. The Molodenski transformation provides the least accuracy.

Exercises

1. Which of the following pairs of parameters is not usually used to define a reference ellipsoid?
   a. The semimajor and semiminor axes
   b. The semimajor axis and the reciprocal of the flattening
   c. The semimajor axis and the eccentricity
   d. The semimajor axis and the reciprocal of the eccentricity

Explanation:

Compared with a sphere, an oblate ellipsoid is flattened at the poles. It is often expressed as a ratio. This ratio, and sometimes its reciprocal, is used as one specification for reference ellipsoids used to model the Earth. It cannot fulfill the definition alone; however, another parameter is always needed. That is frequently the semimajor, the long, and the equatorial axis. An ellipsoid can also be fully determined with the statement of both the semimajor axis and the semiminor, the short, or polar axis. Eccentricity, or first eccentricity, is often used in conjunction with the semimajor axis as the definition of a reference ellipsoid, but the reciprocal of the eccentricity is not.

Answer is (d)

> 2. Which of the following is the reference ellipsoid for NAD27?
>   a. Clarke 1858
>   b. Bessel 1841
>   c. Clarke 1866
>   d. Clarke 1880

Explanation:

In the 1860s while head of the Trigonometrical and Leveling Departments of the Ordnance Survey in Southampton in Britain, Captain Alexander Clarke worked as a geodesist. He made and published calculations of the size and shape of the Earth, first in 1858. Then in 1866 Captain Clarke made his second determination of a reference ellipsoid for the Earth and the figure became a standard reference for U.S. geodesy and the North American Datum 1927. He published another reference ellipsoid in 1880.

Answer is (c)

> 3. Which of the following is the reference ellipsoid for North American Datum 1983?
>   a. OSGB36
>   b. GRS80
>   c. WGS66
>   d. ITRF88

Explanation:

Ordnance Survey Great Britain 1936 or OSGB36 is a traditional geodetic TRF or terrestrial reference frame. Its reference ellipsoid is Airy 1830. The ITRF88 was the first International Terrestrial Reference Frame and its reference ellipsoid is Geodetic Reference System 1980, GRS80, the same used for NAD83.

Answer is (b)

4. Which of the following is not a characteristic of a Conventional Terrestrial Reference System, CTRS?

a. The x, y, and z-axes are all perpendicular to one another

b. The z-axis moves due to polar motion

c. The x and y axes rotate with the Earth around the z-axis

d. The origin is the center of mass of the Earth

Explanation:

All of the answers mention aspects of the conventional terrestrial reference system, CTRS, except (b). While it is true that the actual polar axis wanders, the IERS Reference Pole, IRP, does not move with polar motion because by international agreement it stands where it was at the moment of midnight on the transition between New Year's Eve 1983 and January 1, 1984 (UTC).

Answer is (b)

5. Which of the following statements about GRS80 is not correct?

a. The GRS80 ellipsoid is slightly different from the WGS84 ellipsoid

b. NAD83 uses GRS80 as its reference ellipsoid

c. GRS80 is geocentric

d. The GRS80 ellipsoid is the reference ellipsoid for GPS

Explanation:

The parameters of the Geodetic Reference System 1980, GRS80 adopted by the International Association of Geodesy, IAG during the General Assembly 1979 are as follows:

semiminor axis (polar radius)	B	6356752.3141 m
first eccentricity squared	e^2	0.00669438002290
flattening	F	1 : 298.257222101

The parameters of the World Geodetic System 1984, WGS84, ellipsoid are as follows:

semiminor axis (polar radius)	B	6356752.3142 m
first eccentricity squared	e^2	0.00669437999013
flattening	F	1 : 298.2572223563

These ellipsoids are very similar. The semimajor axis of GRS80 is 6,378,137 m and its inverse of flattening is 298.257222101. The semimajor axis of WGS84 is the same within the accuracy shown at 6,378,137 m, but its inverse of flattening is slightly different 298.257223563.

But GRS80 is not the geodetic reference system used by GPS, that is, the WGS84 ellipsoid. It was developed for the United States Defense Mapping Agency, DMA, now known as the National Geospatial-Intelligence Agency (NGA). GPS receivers compute and store coordinates in terms of WGS84 and most can then transform them to other TRFs.

Answer is (d)

6. Which statement about biaxial and triaxial ellipsoids is correct?

    a. The lengths of the semimajor and semiminor axes are consistent in both biaxial ellipsoid and triaxial ellipsoid

    b. The reference ellipsoid for NAD83 is triaxial and the reference ellipsoid for NAD27 is biaxial

    c. The equator of a triaxial ellipsoid is elliptical; the equator of a biaxial ellipsoid is a circle

    d. Both triaxial and biaxial ellipsoids have flattening at the equator

Explanation:

A triaxial ellipsoid such as the Krassovsky ellipsoid has flattening at the equator but biaxial ellipsoids do not. The equator of a triaxial ellipsoid is elliptical, but the equator of a biaxial ellipsoid is a circle because its semimajor axis has a constant length. GRS80, the reference ellipsoid for NAD83, and the Clarke 1866 ellipsoid, the reference ellipsoid of NAD27, are both biaxial.

Answer is (c)

7. In the Conventional Terrestrial Reference System, CTRS, with its origin at the geocenter of the Earth and a right-handed orientation of the x, y, and z axes the z-coordinate is positive in the northern hemisphere and negative in the southern hemisphere. Which answer correctly states the regions of the Earth where the x-coordinate is positive and where it is negative?

    a. The x-coordinate is negative from the zero meridian to 90° W longitude and from the zero meridian to 90° E longitude. In the remaining 180° of longitude the x-coordinate is positive.

    b. The x-coordinate is positive in the eastern hemisphere and negative in the western.

c. The x-coordinate is positive from the zero meridian to 90° W longitude and from the zero meridian to 90° E longitude. In the remaining 180° the x-coordinate is negative.

d. The x-coordinate is positive in the western hemisphere and negative in the eastern.

Explanation:

In the right-handed three-dimensional coordinate system known as CTRS the x-coordinate is positive from the zero meridian to 90° W longitude and from the zero meridian to 90° E longitude. In the remaining 180° the x-coordinate is negative. The y-coordinate is positive in the eastern hemisphere and negative in the western. The z-coordinate is positive in the northern hemisphere and negative in the southern.

Answer is (c)

8. What constitutes the realization of a geodetic terrestrial reference frame, TRF?

a. The assignment of a semimajor axis and another parameter, such as the first eccentricity, flattening and/or the semiminor axis.

b. The realization on the actual Earth that is provided with known coordinates in the subject TRF.

c. The definition of an initial point at a known latitude and longitude and a geodetic azimuth from there to another known position along with the two parameters of the associated ellipsoid, such as its semimajor and semiminor axes.

d. The transformation of its coordinates into another TR using the seven parameters of the Bursa–Wolf process.

Explanation:

The realization of a TRF is the assignment of coordinates in that TRF, perhaps including actual physical monumentation on the ground. Following the realization of a TRF, subsequent work can be tied to physically marked points on the Earth rather than an abstract concept of the TRF, its theoretical origin, and axes. This concrete manifestation of a TRF is a TRF ready to go to work in the real world.

Answer is (b)

9. Which of the following acronyms identifies a realization of the International Terrestrial Reference System?

    a. VLBI

    b. SLR

    c. ITRF88

    d. IERS

Explanation:

The international organization that compiles the measurements and calculates the movement of our planet is the International Earth Rotation Service or IERS. The first realization of the International Terrestrial Reference System produced by the IERS was the International Terrestrial Reference Frame of 1988 or ITRF88. The IERS also defines the International Reference Pole or IRP. And the data needed for this work comes from GPS, very long baseline interferometry, VLBI, and satellite laser ranging, SLR.

Answer is (c)

10. Which of these transformation methods is likely to produce the most accurate results?

    a. Molodenski

    b. Seven parameter

    c. Surface fitting

    d. Krassovsky

Explanation:

Surface fitting is the best approach to transformation. The United States, Canada, and Australia use the method. Although the number of common points available is an important element in evaluating the accuracy that can be expected from a transformation method, generally the Molodenski transformation provides the least accurate results. The surface-fitting or grid shift techniques like that used in NADCON will provide the most accurate transformation.

Answer is (c)

11. Which of the following statement correctly describes the relationship between positions expressed in the latest realizations of WGS84
and NAD83?

    a. They are the same.

    b. NAD83 positions are virtually identical with ITRF positions, but
WGS84 positions are not.

    c. They are not the same.

    d. NAD83 positions are based on a geocentric system, WGS84 posi-
tions are not.

Explanation:

There is an important global geocentric TRF. The *World Geodetic System 1984*
(WGS84) is the geodetic reference system used by GPS. An attempt was made
to make the GRS80 reference ellipsoid for the new NAD83 geocentric, but the
systems used were inadequate to rigorously determine the center of mass of
the Earth. WGS84 as originally rolled out in 1987 was nearly identical with
NAD83 (1986) and even though WGS84's ellipsoid and the GRS80 ellipsoid
are very similar. The subsequent realizations of WGS84 have caused it to
drift farther and farther from NAD83. Today, the difference between a position in NAD83 (2011) and a position in WGS84 (G1762) can approach 1 or 2 m,
~3 ft to ~6 ft. As it moved away from NAD83, WGS84 has been kept virtually
coincident with the International Terrestrial Reference Frame, ITRF. WGS84
(G1762) positions are virtually identical with ITRF positions, but NAD82
(2011) positions are not. WGS84 (G1762) positions are based on a geocentric
system, NAD83 (2011) positions are not.

Answer is (c)

12. Which of the following does not describe a difference between active
control stations and passive control stations?

    a. Passive control stations are physical monuments typically
embedded in the ground at known locations. Active control stations are also at known locations, but they differ from passive
control stations in that GPS/GNSS receivers are permanently
affixed to each one.

    b. The GPS/GNSS receivers on active control stations collect sat-
ellite positioning data continuously, but such receivers are not
present on passive control stations.

    c. Following the replacement of NAD83 NGS will continue to establish and maintain active control stations but will no longer establish or maintain passive control stations.

    d. Active control stations are constantly moving but passive control stations are not.

Explanation:

Active stations are, like the passive marks, at fixed, official known locations with published coordinates, but they differ in that GPS/GNSS receivers are permanently affixed to each one. The receivers on CORS are continuously collecting satellite positioning data and that data is made available to the public over the Internet. GPS/GNSS signals collected and archived at these sites of known position provide base data that serves as the foundation for the correction and improvement of all sorts of positioning, navigation, and timing work in the United States and its territories. When NAD83 is replaced the new paradigm will also alter the roles and responsibilities that have long supported the NSRS. For many years the NGS has established and maintained the passive control network under the new plan that will end. That responsibility shifts onto the users. If users need passive control monuments they will need set, and tie them to the NSRS themselves, but such control will not be considered an official part of the NSRS.

In the past official, the static coordinates published by the NGS for passive control stations were assumed to be unchanging. The measurement systems available at the time created sufficiently reliable positions on passive monuments. Their constant slow movement was not apparent then because the errors in the measurement systems exceeded the extent of the motion. That is no longer the case. Measurement technology has improved dramatically. It now reveals the persistent movement.

Answer is (d)

3

Heights

Latitude and longitude, northing and easting, radius vector and polar angle, and coordinates often come in pairs. But that is not the whole story. For a coordinate to entirely describe the point it represents must lie on a well-defined surface. It might be a flat plane or the surface of a particular ellipsoid and in either case the surface will be smooth and have a definite and complete mathematical definition.

As mentioned before, today geodetic reference frames rely on the surfaces of ellipsoids to approximate the surface of the Earth. But the actual Earth does not coincide with these nice smooth surfaces, even though that is where the points represented by the coordinate pairs lay. In other words, the abstract points are on the ellipsoid, but the physical features are, of course, on the actual Earth. Though the intention is for the Earth and the ellipsoid to have the same center, the surfaces of the two figures are certainly not in the same place. There is a distance between them.

Ellipsoid Height

The distance represented by a coordinate pair on the reference ellipsoid to the point on the surface of the Earth is measured along a line perpendicular to the ellipsoid. This distance is known by more than one name. It is called the *ellipsoidal height* and it is also called the *geodetic height* and is usually symbolized by h.

In Figure 3.1 the ellipsoidal height of station Youghall is illustrated. The reference ellipsoid is GRS80.

The concept of an ellipsoidal height is straightforward. The surface of a reference ellipsoid may be above or below the surface of the Earth at a particular place. If the ellipsoid's surface is below the surface of the Earth at the point, the ellipsoidal height has a positive sign, if the ellipsoid's surface is above the surface of the Earth at the point, the ellipsoidal height has a negative sign. It is important to remember that the measurement of an ellipsoidal height is along a line perpendicular to the ellipsoid, not along a plumb line. Said another way, an ellipsoidal height is not measured in the direction of gravity. It is not measured in the conventional sense of down or up. As was mentioned in Chapter 1, in the case of an ellipsoidal height, down is a line

FIGURE 3.1
Ellipsoidal height.

perpendicular to the ellipsoidal surface at a particular point on the ellipsoidal model of the Earth.

However, on the real surface of the Earth, down is typically considered to be in the direction of gravity at the point. Most often they are not the same. And since a reference ellipsoid is a geometric imagining, it is quite impossible to actually set up an instrument on it. That makes it tough to measure ellipsoidal height using a conventional surveying instrument. In other words, ellipsoidal height is not what most people think of as an elevation.

The ellipsoidal height of a point is readily determined using a GPS/GNSS receiver. GPS/GNSS can be used to discover the distance from the geocenter of the Earth to any point on the Earth, or above it for that matter. In other words, it has the capability of determining three-dimensional coordinates of a point in a short time. It can provide latitude, longitude, and if the system has the parameters of the reference ellipsoid in its software it can calculate the ellipsoidal height. The relationship between points can be further expressed in the ECEF coordinates, x, y, and z, or in a *Local Geodetic Horizon System*, *LGHS* also abbreviated *LGH*, of north, N, east, E, and up, U, as shown in Figure 3.2. The up, U, can be an ellipsoidal height in LGH.

In a manner of speaking, ellipsoidal heights are new, at least in common usage, because they could not be easily determined until GPS/GNSS became

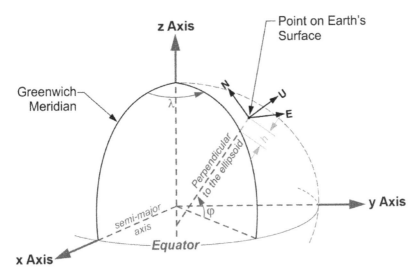

FIGURE 3.2
Local Geodetic Horizon System, LGHS.

a practical tool in the 1980s. However, ellipsoidal heights are not all the same, because reference ellipsoids or sometimes just their centers can differ. When the ellipsoid changes, the h changes too.

There is nothing new about heights themselves or elevations as they are often called. Long before ellipsoidal heights were so conveniently available, knowing the elevation of a point was critical to the complete definition of a position. In fact there are more than 200 different vertical reference frames in use in the world today. They were, and still are, determined by a method of measurement known as *leveling*. But it is important to note that this process measures a height that is not ellipsoidal.

Both *trigonometric leveling* and *spirit leveling* depend on optical instruments. Their lines of sight are oriented to gravity not a reference ellipsoid.

Trigonometric Leveling

Finding differences in heights with trigonometric leveling requires a level optical to measure angles in the vertical plane, a graduated rod, and either a known horizontal distance or slope distance between the two of them. As shown in Figure 3.3 the instrument is centered over a point of known elevation and the rod is held vertically on the point of unknown elevation. In the instrument, one of the two angles is measured, either the vertical angle, from horizontal plane of the instrument, or the zenith angle, from the instrument's

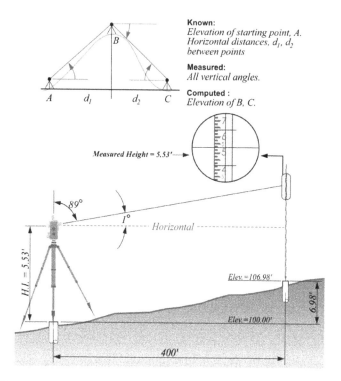

FIGURE 3.3
Trigonometric leveling.

vertical axis. Either angle will do. This measured angle together with the distance between the instrument and the rod provides two known components of right triangle in the vertical plane. It is then possible to solve that triangle to reveal the vertical distance between the point at the instrument and the point on which the rod is held.

For example, suppose that the height, or elevation, of the point over which the instrument is centered is 100.00 ft. Further, suppose that the height of the instrument's level line of sight, its horizontal plane is 5.53 ft above that point. Then the *height of the instrument (H.I.)* would then be 105.53 ft. For convenience, the vertical angle at the instrument could be measured to 5.53 ft on the rod. If the measured angle is 1° 00′ 00″ and the horizontal distance from the instrument to the rod is known to be 400.00 ft, all the elements are in place to calculate a new height. In this case the tangent of 1° 00′00″ multiplied by 400.00 ft yields 6.98 ft. That is, the difference in height from the point at the instrument and the point at the rod. Therefore, 100.00 ft plus 6.98 ft indicates a height of 106.98 ft at the new station where the rod was placed.

There are many more aspects to this process, the curvature of the Earth, refraction of light, and so on, that make it much more complex in practice than it is in this illustration. However, the fundamental of the procedure is the solution of a right triangle in a vertical plane using trigonometry, hence the name trigonometric leveling. It is faster and more efficient than spirit leveling but it is generally not as precise.

Horizontal surveying often precedes leveling in control networks. That was true in the early days of what has become our national network, the National Spatial Reference System, NSRS, of the United States. Geodetic leveling was begun only after triangulation networks were underway. This was also the case in many other countries. In some places around the world the horizontal work was even completed before leveling was commenced. In the United States, trigonometric leveling was applied to geodetic surveying before spirit leveling. Trigonometric leveling was used extensively to provide elevations to reduce the angle observations and base lines necessary to complete triangulation networks to a version of *sea level*. And the angular measurements for the trigonometric leveling were frequently done in an independent operation with instruments having only a vertical circle.

Then in 1871 Congress authorized a change for the then *Coast Survey* under Benjamin Peirce that brought spirit leveling to the forefront. The Coast Survey was to begin a transcontinental arc of triangulation to connect the surveys on the Atlantic coast with those on the Pacific coast. Until that time their work had been restricted to the coasts. But with the undertaking of triangulation that would cross the continent along the 39th parallel it was clear that trigonometric leveling was not sufficient to support the project. They needed more vertical accuracy than it could provide. So in 1878 at about the time the work began the name of the agency was changed from Coast Survey to U.S. Coast and Geodetic Survey, and a line of spirit leveling of high precision was begun at *Benchmark A* in the foundation wall of the Washington County Court House in Hagerstown, Maryland. It headed west. It reached Seattle in 1907. Along the way it provided passive monumented benchmarks for the use of engineers and others who needed accurate elevations and heights, for subsequent work not to mention establishing the vertical datum for the United States.

Spirit Leveling

The method is simple in principle, but not in practice. An instrument called a level is used to establish a line of sight that is perpendicular to gravity, in other words, a level line. Then two rods marked with exactly the same graduations, like rulers, are held vertically resting on two solid points, one ahead and one behind the level along the route of the survey. As shown in Figure 3.4 the system works best when the level is midway between these rods. Looking at the rod to the rear through the telescope of the level, the point at which the horizontal level line of sight of the level intersects the vertical rod there is a

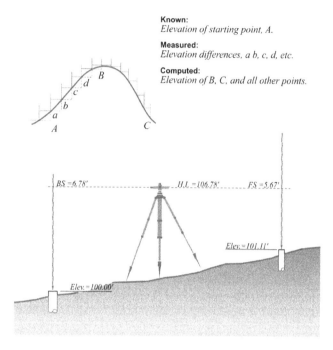

Known:
Elevation of starting point, A.

Measured:
Elevation differences, a b, c, d, etc.

Computed:
Elevation of B, C, and all other points.

BS = 6.78'   *H.I. = 106.78'*   *FS = 5.67'*

Elev. = 101.11'

Elev. = 100.00'

FIGURE 3.4
Spirit leveling.

graduation. That reading is taken and noted. This is known as the *backsight* (B.S.). This reading tells the height, or elevation, that the level line of sight passes above the mark on which the rod is resting. For example, if the point on which the rod is resting is at an elevation of 100 ft and the reading on the rod is 6.78 ft then the height of the level's line of sight is 106.78 ft. That value is known as the H.I. Then the still level instrument is rotated to observe the vertical rod ahead and a value is read there. This is known as the *foresight* (F.S.). The difference between the two readings reveals the change in elevation from the first point at the backsight to the second, at the foresight. For example, if the first reading established the height of the level's line of sight, the H.I., at 106.78 ft, and the reading on the rod ahead, the F.S., was 5.67 ft it becomes clear that the second mark is 1.11 ft higher than the first. It has an elevation of 101.11 ft. By beginning this process from a monumented point of known height, a benchmark, and repeating it with good procedures the heights of marks can be determined all along the route of the survey.

The accuracy of level work depends on the techniques and the care used. Methods such as balancing the forward and back sights, calculating refraction errors, running new circuits twice, using one piece rods, and so on, can improve results markedly. In fact, entire book has been written on the details of proper leveling techniques. Here the goal will be to mention just a few pertinent elements.

It is difficult to overstate the amount of effort devoted to differential spirit level work that has carried vertical control and established passive monument benchmarks across the United States. The transcontinental precision leveling surveys done by the Coast and Geodetic Survey from coast-to-coast were followed by thousands of miles of spirit leveling work of varying precision. When the 39th parallel survey reached the west coast in 1907 there were approximately 23,835 miles, 38,359 km, of geodetic leveling in the national network. That was more than doubled 22 years later in 1929 to approximately 46,700 miles, 75,159 km in the United States. Inevitably, this growth in leveling information and benchmarks made a new general adjustment of the network necessary. In part because as the quantity of leveling information grew, so did the errors and inconsistencies much of it due to the accumulation of error inherent in conventional surveying, as was mentioned in Chapter 2, but there were other causes too. For example, the foundation of the work was ultimately intended to be *Mean Sea Level, MSL* as measured by *tide station gauges* and that presented a problem.

There had already been four previous general adjustments to the vertical network across the United States by 1929. They were done in 1900, 1903, 1907, and 1912. The adjustment in 1900 was based on elevations held to the Mean Sea Level as determined at five tide stations. The adjustments in 1907 and 1912 left the eastern half of the United States fixed as adjusted in 1903. In 1927 there was a special adjustment of the leveling network. This adjustment was not fixed to Mean Sea Level at all tide stations and after it was completed, it became apparent that the Mean Sea Level surface as defined by tidal observations had a tendency to slope upward to the north along both the Pacific and Atlantic Coasts, with the Pacific being higher than the Atlantic. However, in the adjustment that established the *Sea Level Datum of 1929*, the determinations of Mean Sea Level at 26 tide stations, 21 in the United States and 5 in Canada, were held fixed. Since Mean Sea level was the intended foundation of these adjustments, and it might make sense to say a few words about the forces that shape it.

Sea Level

All celestial bodies, but particularly the Sun and the Moon exert tidal forces that affect the waters and atmosphere of the Earth as well as the body of the Earth itself. The Moon's force is greater of the two. The Sun's tidal force is about half of that exerted on the Earth by the Moon. The Moon makes a complete elliptical orbit around the Earth every 27.3 days. There is a gravitational force between the Moon and the Earth. Each pulls on the other. And at any particular moment the gravitational pull is greatest on the portion of the Earth that happens to be closest to the Moon. That produces a bulge in the waters on the Earth in response to the tidal force. On the side of the Earth opposite the bulge centrifugal force exceeds the gravitational force of the Earth, and water in this area is forced out away from the surface of the Earth creating another bulge. But the two bulges are not stationary, they move across the surface of the Earth. They move because not only is the Moon moving slowly relative to the Earth as it proceeds along its orbit, but more importantly the Earth is rotating in relation to the Moon. And the Earth's rotation is relatively rapid in comparison with the Moon's movement. Therefore, a coastal area in the high middle latitudes may find itself with a high tide early in the day when it is close to the Moon, a low tide in the middle of the day when it has rotated away from it. And this cycle will begin again with another high tide a bit more than 24 h after the first high tide. A bit more than 24 h because from the moment the Moon reaches a particular meridian to the next time there is actually about 24 h and 50 min, a period is called a *lunar day*.

This sort of tide with one high water and one low water in a lunar day is known as a *diurnal tide*. This characteristic tide would be most likely to occur in the middle latitudes to the high latitudes when the Moon is near its maximum declination as you can see from Figure 3.5.

The declination of a celestial body is similar to the latitude of a point on the Earth. It is an angle measured at the center of the Earth from the plane of the equator, positive to the north and negative to the south, to the subject, which is in this case the Moon. The Moon's declination varies from its minimum of 0° at the equator to its maximum over a 27.2-day period, and that maximum declination oscillates too. It goes from ±18.5° up to ±28.5° over the course of an 18.6-year cycle.

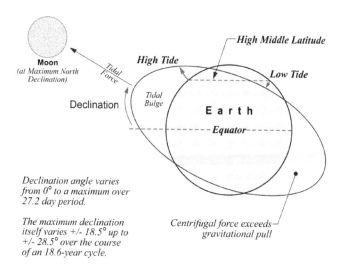

FIGURE 3.5
Diurnal tide.

Another factor that contributes to the behavior of tides is the elliptical nature of the Moon's orbit around the Earth. When the Moon is closest to the Earth, that is its *perigee*, the gravitational force between the Earth and the Moon is 20% greater than usual. At *apogee*, when the Moon is farthest from the Earth the force is 20% less than usual. The variations in the force have exactly the affect you would expect on the tides, making them higher and lower than usual. It is about 27.5 days from perigee-to-perigee.

To summarize, the Moon's orbital period is 27.3 days. It also takes 27.2 days for the Moon to move from its maximum declinations back to 0° directly over the equator. And there are 27.5 days from one perigee to the next. You can see that these cycles are almost the same, almost, but not quite. They are just different enough that it takes from 18.6 years for the Moon to go through all of the possible combinations of its cycles. And therefore, if you want to be certain that you have recorded the full range of tidal variation at a place you must observe and record the tides at that location for 19 years. Nineteen years is usually quoted because it is the smallest integer number of years larger than 18.6.

This 19-year period, sometimes called the *Metonic* cycle, is the foundation of the definition of Mean Sea Level. Mean Sea level, MSL, can be defined as the arithmetic mean of hourly heights of the sea at a primary-control tide station observed over a period of 19 years. The mean in Mean Sea Level refers to the average of these observations over time at one place. It is important to note that it does not refer to an average calculation made from measurements at several different places. Therefore, when the Sea Level Datum of 1929 was fixed to MSL at 26 tide stations that meant it was made to fit 26 different and distinct *Local Mean Sea Levels*, LMSL. In other words, it was warped to coincide with 26 different elevations.

The topography of the sea changes from place-to-place and this means, for example, that MSL in Florida is not the same as MSL in California. The fact is Mean Sea Level varies. The water's temperature, salinity, currents, density, wind, and other physical forces all cause changes in the sea surface's topography. For example, the Atlantic Ocean north of the Gulf Stream's strong current is around 1 m, 3.3 ft, lower than it is further south. The more dense water of the Atlantic is generally about 40 cm, 1.3 ft, lower than the Pacific. At the Panama Canal the actual difference is about 20 cm, 0.66 ft, from the east end to the west end. The *Local Mean Sea Level*, LMSL, at the San Francisco tidal station differs from today's national vertical control value thereby nearly 1 m, 3.3 ft.

Evolution of the Vertical Datum

After it was formally established, thousands of miles of leveling were added to the Sea Level Datum of 1929, *SLD29*. The Canadian network also contributed data to the Sea Level Datum of 1929, but Canada did not ultimately use what became known as the *National Geodetic Vertical Datum of 1929, NGVD 29.* The name was changed from SLD29 to NGVD29 in 1976. It did not really coincide with Mean Sea Level. The precise leveling done to produce the fundamental data was internally consistent, but when the network was warped to fit so many tide station determinations of Mean Sea Level consistency suffered.

By the time the name was changed to NGVD29 there were approximately 400,000 miles of new leveling work and about 51,000 miles of first-order releveling. There was also a more complete understanding of the differences in the Local Mean Sea Levels at the previously mentioned tidal gauges, new leveling corrections, and terrestrial gravity measurements. This new knowledge revealed distortions in the network. There was also difficulty with the user's access to NGVD29. Many of the original passive benchmarks had been disturbed, destroyed, or lost. In the 1970s the National Geodetic Survey (NGS) thought that it is time for a renewed effort and a new adjustment of the North American vertical datum.

The Zero Point

This time there was a different approach. Instead of fixing the adjustment to many tidal stations the new adjustment would be minimally constrained. It would be fixed to only one station. There was precedence for the idea. The tide gauge at Pointe-au-Père, Quebec, near the mouth of the St. Lawrence River and on its southern bank, had already been serving as the single zero point for a vertical datum known as the *International Great Lakes Datum 1955, IGLD 1955.* That datum provided the foundation for monitoring the water level in the Great Lakes.

The International Great Lakes Datum

In 1953 well before the new adjustment for the replacement of NGVD29 had begun the IGLD precise leveling proceeded from the zero reference on a tide gauge established at Pointe-au-Père, Quebec. In September of 1961, published elevations for the monumented benchmarks were released and IGLD 1955 was realized.

In the years that followed a few events necessitated an update to IGLD 1955. One was the relocation of the Pointe-au-Père, Quebec tide gauge. The wharf at Pointe-au-Père had deteriorated, so the zero point was transferred to Father Point/Rimouski, Quebec 3 miles away. Another was *isostatic rebound,* this effect is literally the Earth's crust rising slowly, rebounding, from the removal of the weight and subsurface fluids caused by the retreat of the glaciers from the last ice age. The uplift changed the heights of the previously established passive benchmarks.

Therefore, beginning in 1976 several agencies including the *National Oceanic and Atmospheric Administration, NOAA,* the United States Army Corps of Engineers, the *Canadian Hydrographic Service,* and the *Geodetic Survey of Canada* worked together to realize the *International Great Lakes Datum 1985, IGLD 1985.*

The North American Vertical Datum 1988

Father Point/Rimouski, Quebec also became the zero point for the new adjustment of the North American vertical datum known as the *North American Vertical Datum 1988, NAVD88.* This step ensured that the zero reference surface would be somewhere near MSL. However, the foremost reasons for the decision were to provide agreement with the International Great Lakes Datum of 1985 and to lessen the recompilation efforts required to update existing national mapping products including the United States Geological Survey, USGS, topographic map series, a significant consideration when the mapping existed only on paper. Fixing to a single point also eliminated the warping inherent in NGVD29 caused by its constraint to dozens of tidal gauges. Unfortunately, it also led to the accumulation of systematic error as the geodetic leveling spread out from Father Point/Rimouski. 81,500 km (50,600 miles) of releveling began in the 1970s; it addressed the elevations of benchmarks all across the nation. Destroyed and disturbed benchmarks were replaced. The new adjustment involved approximately 505,000 permanent benchmarks, 204,000 temporary benchmarks, and 1.2 million observations.

NAVD88 was ready in June of 1991. In the 48 conterminous states of the United States the differences between elevations of benchmarks determined in NGVD29 compared with the elevations of the same benchmarks in NAVD88 vary from approximately −1.3 ft in the east to approximately +4.9 ft in the west. The larger differences tend to be on the coasts, as one would expect since NGVD29 was forced to fit *Mean Sea Level, MSL,* at many tidal stations and NAVD88 was held to just one. In Alaska the variation is approximately +3.0 ft to approximately +7.9 ft.

When comparing heights in IGLD85 and NAVD88 it is important to consider that even though they are both based on the same zero point there is a difference between the natures of the heights in the two systems. NAVD88 values are expressed in *Helmert orthometric height* units and IGLD85 elevations are given in *dynamic height* units. The explanation of this difference requires introduction of some additional principles.

So far there has been mention of heights based on the ellipsoidal model of the Earth and heights that use Mean Sea Level as their foundation and the fact that Mean Sea Level is affected by a broad range of physical forces. There is another pertinent surface to which heights are referenced. That surface is defined by one force, that is, gravity. It is known as a geoid.

Geoid

Any object in the Earth's gravitational field derives potential energy from being pulled downward and there is a relationship between an object's height and the amount of that potential energy. Here is another way of saying the same thing. The potential energy an object derives from gravity is the same as the work it would take to lift its height. Imagine several objects, each with the same weight resting on a truly level floor. They would all be possessed of the same potential energy from gravity. Their potential energies would be equal. The floor on which they were resting could be said to be a surface of equal potential, an *equipotential* surface. Now suppose that each of the objects was lifted up onto a level table. Their potential energies would obviously be changed in the process. Once they were all resting on the table their potential energies would again be equal, now on a higher equipotential surface, but how much higher?

There is more than one way to answer that question. One way is to find the difference in their *geopotential* that resulted from the lifting. What is their potential energy on the floor, thanks to gravity, compared with their geopotential on the table? This is the same idea behind answering with a dynamic height as is done in IGLD85. Another way to answer the question is to simply measure the distance along a plumb line from the floor to the tabletop. This latter method is the basic idea behind an orthometric height. In other words, orthometric heights are geometric. An orthometric height can be further explained by imagining that the floor in the example is a portion of one particular equipotential surface.

There are an infinite number of equipotential surfaces. However, Mean Sea Level is not one of them. It is not a surface on which the geopotential is always the same. It could be so if the oceans of the world could be utterly still, completely free of currents, tides, friction, variations in temperature, and all other physical forces, except gravity. If the water was then directed by small frictionless channels or tubes and allowed to migrate across the land, the water would then, theoretically, define an equipotential surface across the continents. Of course, 70% of the earth covered by oceans is not so cooperative, and the physical forces such as temperature, salinity, currents, wind, and so on cannot really be eliminated. These unavoidable forces actually cause Mean Sea Level to deviate up to 1 m (3.3 ft), even 2 m (6.6 ft) from an equipotential surface.

On the other hand, a geoid is an equipotential surface because it is defined by gravity alone. It can follow the particular equipotential surface that best fits Mean Sea Level, in a least squares sense (see Figure 3.6), or it can follow the equipotential surface that best accommodates a national vertical datum. In either case along the surface of a geoid the potential of gravity, the geopotential, is consistent.

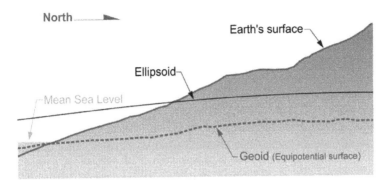

FIGURE 3.6
Ellipsoid and geoid.

As a geoid is completely defined by gravity it is not smooth. It is an undulating convex surface. As shown in the exaggerated illustration Figure 3.7. A geoid is irregular because gravity is not consistent across the surface of the Earth. It varies with the uneven distribution of the mass within the earth. It has all the irregularity that the attendant variation in gravity implies. In fact, the separation between a geoid and the smooth GRS80 ellipsoid worldwide varies from about +85 m, +279 ft, west of Ireland to about −106 m, −348 ft, and the latter in the area south of India near Ceylon.

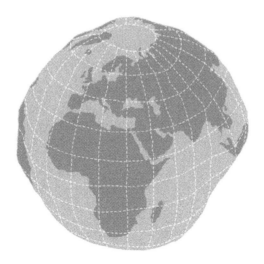

FIGURE 3.7
Exaggerated representation of a geoid.

Dynamic Heights

Now please recall that an orthometric height is a measurement along a plumb line from a particular equipotential surface to a point on the Earth's surface. In other words, the orthometric height of that point is its distance from the reference surface, a distance that is measured along the line perpendicular to every equipotential surface in between. And these equipotential surfaces are not parallel with each other, chiefly because of gravity anomalies and the rotation and shape of the Earth. Therefore, it follows that two points could actually have the same orthometric height and not be on the same equipotential surface. A rather odd fact and it has an unusual implication. It means that water might actually flow between two points that have exactly the same orthometric height.

This is one reason that the International Great Lakes Datum of 1985 is based on dynamic heights. Unlike orthometric heights, two points with identical dynamic heights are definitely on the same equipotential surface. Two points would have to have different dynamic heights for water to flow between them. And the flow of water is a critical concern for those using that system. As one can rely that points with the same dynamic heights are always on the same equipotential surface they are better indicators of the behavior of water than orthometric heights.

An advantage of using dynamic heights is the sure indication of whether a water surface, or any other surface for that matter, is truly level or not. For example, the Great Lakes are monitored by tide gauges to track historical and predict future water levels in the lakes, and it is no surprise then that the subsequent International Great Lakes Datum of 1985, IGLD85, heights are expressed as dynamic heights. Where the lake surfaces are level the dynamic heights are the same, and where they are not it is immediately apparent because the dynamic heights differ.

Points on the same equipotential surface also have the same geopotential numbers along with the same dynamic heights. The idea of measuring geopotential by using geopotential numbers was adopted by the International Association of Geodesy in 1955. The geopotential number of a point is the difference between the geopotential below the point, down on a geoid, and the geopotential right at the point itself. It is positive because the potential decreases linearly with the distance away from the Earth. Said another way, the geopotential number expresses the work done if a weight is lifted from a geoid up to the point, like the weights that were lifted onto a table in the earlier analogy. Also, geopotential heights differ from heights derived from uncorrected differential leveling circuit. The uncorrected orthometric heights from leveling are somewhat dependent on the route taken by the circuit. That method does not produce a single-value answer. However, a geopotential height is a single value answer because the geometric height differences observed by differential leveling are added to a correction term that accounts for gravity.

A geopotential number is expressed in geopotential units, or gpu. A gpu is 1 kgal/m, or 1000 gal/m. Geopotential numbers along with a constant contribute the calculation of dynamic heights. The calculation itself is simple.

$$H_p^{dyn} = \frac{C_p}{\gamma_0}$$

where:

H_p^{dyn} is the dynamic height of a point in meters

C_p is the geopotential number at that point in gpu, that is kilogals per meter

γ_0 is the constant 0.9806199 kgals

The constant is normal gravity at 45° latitude on GRS80.

A dynamic height of a point can be found by dividing its NAVD88 geopotential number by the normal gravity value. In other words, dynamic heights can be geopotential numbers scaled by a particular constant value chosen in 1984 to be normal gravity at 45° latitude on the GRS80 reference ellipsoid. The whole point of the calculation is to transform the geopotential number, that is, in kilogals per meter into a dynamic height in meters by dividing by the constant, that is, in kilogals. Here is an example calculation of the dynamic height of station M 393, PID KK1340, an NGS benchmark.

$$H^{dyn} = \frac{C}{\gamma_0}$$

$$H^{dyn} = \frac{1660.419936 \text{ gpu}}{0.9806199 \text{ kgals}}$$

$$H^{dyn} = 1693.235 \text{ m}(5555.23 \text{ ft})$$

The NAVD88 orthometric height of this benchmark determined by spirit leveling is 5560.79 ft, 1694.931 m, and differs from its calculated dynamic height by 5.56 ft, 1.696 m. However, using the same formula and the same geopotential number, but divided by a gravity value derived from the Helmert height reduction formula, the result would be a Helmert orthometric height for M 393, instead of its dynamic height. Note that the geopotential number stays the same in both systems. And there is a third, if the divisor were a gravity value calculated with the international formula for normal gravity, the answer would be the normal orthometric height for the point.

Measuring Gravity

At every point gravity has a magnitude and a direction. Anywhere on the Earth, a vector can describe gravity, but these vectors do not all have the *same* direction or magnitude. Some parts of the Earth are denser than others and the Earth is not a sphere where both affect gravity. It follows then that defining a geoid precisely involves actually measuring the direction and magnitude of gravity everywhere, but how?

It is quite difficult to gather sufficient data to map a geoid well. Perfection is unlikely to be achieved, but improvements have been accomplished with regularity by using gravity measurements from several different sources. Some gravity measurements are done with a class of instruments called *gravimeters*, which were introduced in the middle of the last century. One sort of gravimeter can be used to measure the relative difference in the force of gravity from place-to-place. The basic idea of this kind of gravimeter is illustrated by considering a weight suspended at the end of a spring. Suppose the extension of the spring was carefully measured with the gravimeter on the table and measured again when it was on the floor. If the measurement of such a tiny difference were possible the spring would be found to be infinitesimally longer on the floor because the magnitude of gravity increases as you move lower. In practice such a measurement is quite difficult, so it is the increase in the tension on the spring necessary to bring the weight back to a predefined zero point that is actually measured. Conversely, suppose the tension of the spring was carefully measured with the gravimeter on the floor and measured again on the table, the spring would be shortened and the tension on the spring would need to decrease to bring the weight back to the zero point. This is because the spinning of the Earth on its axis creates a *centrifugal* aka center fleeing force.

There are two forces to consider here: the centrifugal force and the downward gravitational attraction. These two forces oppose one another and they are indistinguishably bound up with each other. Therefore, measurements made by gravimeters on the Earth inevitably contain both centrifugal and gravitational forces. It is impossible to pry them apart. An idea called the *Equivalence Principle* states that the effects of being accelerated to a velocity are indistinguishable from the effects of being in a gravitational field. In other words there is no physical difference between an accelerating frame of reference and the same frame of reference in a gravitational field. So they are taken together. In any case, as you go higher the centrifugal force increases and counteracts the gravitational attraction to a greater degree than it does at lower heights and so, generally speaking, gravity decreases the higher you go.

Now please recall that the Earth closely resembles an oblate spheroid. That means that the distance from the center of mass of the planet to a point on the equator is longer than the distance from the center to the poles. Said another way, the Earth is generally higher at the equator than it is at the poles. As a consequence the acceleration of a falling object is less at the equator than at the poles. Less *acceleration of gravity* means that if you drop a ball at the equator the rate at which its fall would accelerate would be less than if you dropped it at one of the poles. As a matter of fact that describes the basic idea behind another kind of gravimeter. In this second kind of gravimeter the fall of an object inside a vacuum chamber is very carefully measured.

Terrestrial gravity measurements, measurements from Earth orbiting satellites, and aircraft and ships are all used together to produce a model of the Earth's gravitational field. Large areas can be covered most efficiently with satellites such as the NASA Gravity Recovery and Climate Experiment, GRACE and the European Space Agency's Gravity Field and Steady-State Ocean Circulation (GOCE) satellite. Both were launched from the Plesetsk Cosmodrome, Russia on Rocket launch vehicles, GRACE in 2002 and GOCE in 2009. GOCE's mission was completed in 4 years and 8 months. GRACE provided the first global coverage from a single source. It employs two satellites flying close together in the same orbital plane. The distance between them is constantly being measured using a microwave system. When the lead satellite passes over an anomaly with stronger gravity it pulls away somewhat from its partner satellite. When the trailing satellite passes over the same place the change in gravity makes it advance a bit toward the leader and GRACE's. GPS measurements pinpoint the spot. Any such movement that is not caused by gravitational forces can be eliminated from the data by measurements from the onboard accelerometers. The system can produce monthly gravity mapping with a 100-fold increase in precision over earlier systems. When the GRACE mission ends a follow-on mission is planned, GRACE-FO.

However, even the best satellite-based missions cannot deliver the details of gravity that terrestrial and airborne gravity mapping can supply. Therefore, the NGS program known as Gravity for the Redefinition of the American Vertical Datum (GRAV-D) utilizes a variety of aircraft that fly patterns over the areas of interest with a Turn-key Airborne Gravimetry System, TAGS, on board. TAGS utilizes a spring gravity sensor on a gyro-stabilized platform that has the capability to measure gravity to a precision of 0.01 milligal. Ok, what is a gal?

Gal

Gravity is quantified in a unit of measurement known as a gal. It was named for Galileo, who pioneered the modern understanding of gravity. Imagine an object is dropped from a high ledge. At the end of the first second it is falling at 1 cm (0.33 ft) per second. Then at the end of the next second it is falling at 2 cm (0.66 ft) per second. The imagined object would have just accelerated 1 gal. 1 gal is an acceleration of 1 cm (0.33 ft) per second per second and the acceleration of gravity, the rate at which the velocity of a falling object changes, is expressed in gals.

At the equator the average acceleration of a falling object is approximately 978 gals, that is, 978 cm/s² or 32.09 ft/s². At the poles the acceleration of a falling object increases to approximately 984 gals, which is 984 cm/s² or 32.28 ft/s². The acceleration of a falling object at 45° latitude is between these two values, as you would expect. It is 980.6199 gals, 980.6199 cm/s², or 32.17 ft/s². This value is called *normal gravity*.

Orthometric Correction

Imagine the equipotential surfaces that surround the Earth as the layers of an onion. These layers are farther apart at the equator than they are at the poles. In other words, equipotential surfaces get closer together as you approach the poles, they converge. And the effect of this convergence becomes more pronounced as the direction of your route gets closer to cardinal north or south.

This effect was mentioned as far back as 1899. It was discovered that the precise leveling run in an east–west direction required less correction than leveling done in a north or south direction. Eventually, a value known as an *orthometric correction* was applied to accommodate the convergence of the equipotential surfaces. As Howard Rappleye wrote,

> The instruments and methods used in 1878 were continued in use until 1899, when, as the result of an elaborate theoretical investigation, it was found that apparently the leveling was subject to a systematic error depending on the azimuth of the line of levels... The leveling of the Coast and Geodetic Survey was then corrected for the systematic error.

(Rappleye 1948, p. 1)

In the years that led up to the establishment of NGVD29 gravity data was mostly unavailable and the actual correction applied was based on an ellipsoidal model. The result is known as *normal orthometric heights* and do not take account of local variations in gravity. The application of the *orthometric correction* meant that the height difference derived from leveling between two points would not exactly match the difference between orthometric heights. This is a consequence of the fact that the line of sight of a properly balanced level follows an equipotential surface, that is a level surface, but orthometric heights are geometric and after their correction are not exactly on a level surface.

Consider two points, two benchmarks, with the same published ortho-metric height, one north and one south. The orthometric height of the south benchmark as measured along a plumb line from a geoid will pass through fewer level surfaces than the same measurement to the benchmark at the north end. For example, the orthometric height of the water surface at the south end of Lake Huron seems to indicate that it is approximately 5 cm higher than the same equipotential surface at the north end. The orthometric heights make it look this way because the equipotential surfaces are closer together at the north end than they are at the south end. But precise levels run from the south end of the lake to the north would not reflect the 5 cm difference because the line of sight of the level would actually follow the equipotential surface all the way.

In other words, the convergence of equipotential surfaces as shown in Figure 3.8 prevents leveling from providing the differences between points as it is defined in orthometric heights. And the amount of the effect depends on the direction of the level circuit. It is sometimes said that uncorrected dif-ferential leveling does not produce a *single-value* answer because the result is somewhat dependent on the route taken by the circuit. The problem can be alleviated somewhat by applying an orthometric correction based on the measurement of gravity that can amount to 0.04 ft per mile in mountainous areas.

It is worthwhile to note that NGS publishes their height data as Helmert orthometric heights. This is a particular type of orthometric height that does not take account of the gravitational effect of topographic relief. As a con-sequence they can lead to a certain level of misclosure between GPS/GNSS determined benchmarks and a geoid model in mountainous areas.

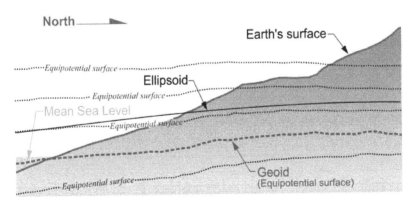

FIGURE 3.8
Equipotential surfaces converging.

Ellipsoid, Geoid, and Orthometric Heights

The distance measured along a line perpendicular to the ellipsoid from the ellipsoid of reference to a geoid is known as a *geoid height*. It is usually symbolized, N, as shown in Figure 3.9. In the CONUS geoid heights vary from about −8 m, −26 ft, to about −53 m, −174 ft, in NAD83. If the geoid is above the ellipsoid N is positive, if the geoid is below the ellipsoid N is negative. Throughout CONUS it is underneath the ellipsoid and N is negative. In Alaska it is the other way around, the ellipsoid is underneath the geoid and N is positive.

Also shown in Figure 3.9 both ellipsoidal and orthometric heights are measured to a point on the surface of the Earth. An ellipsoid height is symbolized, h, is along a line perpendicular to the ellipsoid of reference. An orthometric

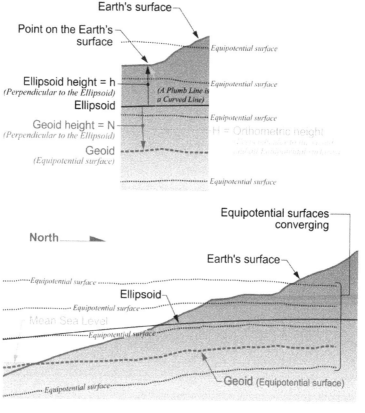

FIGURE 3.9
Orthometric height (H), ellipsoidal height (h), and geoid height (N).

height, symbolized, H, is measured along a plumb line aligned with gravity. In either case by using the formula,

$$H = h - N$$

One can convert an ellipsoidal height, h, derived from a GPS/GNSS observation, into an orthometric height, H, by knowing the extent of geoid-ellipsoid separation, a geoid height, N, at that point. That separation depends on the location of the geoid, of course. As you can see from Figure 3.9 the ellipsoid height of a particular point is actually smaller than the orthometric height throughout CONUS.

However, it might be best to express the formula in this way $H \approx h - N$ because it does not account for all of the elements involved in the calculation. The effects of some are small. For example, it does not account for the fact that the plumb line along which an orthometric height is measured is curved. It is curved because it is perpendicular with each and every equipotential surface through which it passes. And since equipotential surfaces are not parallel with each other, the plumb line must be curved to maintain perpendicularity with all of them. This deviation of a plumb line from the perpendicular to the ellipsoid reaches about 1 min of arc in only the most extreme cases. Therefore, any height difference that is caused by the curvature is negligible. Conversely, the effects of the systematic and random errors in the ellipsoidal height measurements by GPS/GNSS, geodetic leveling, and gravity measurements used to create the geoid models are more significant.

NGS Geoid Models

The shape of a geoid defies the certain, clear definition of, say, the GRS80 ellipsoid. It does not precisely follow Mean Sea Level, and neither does it exactly correspond with the topography of the dry land. It is irregular like the terrestrial surface. The variation is due to the uneven distribution of the mass of the planet. The undulations of a geoid are called *gravity anomalies*. Gravity anomalies are quantified by the difference between the actual measured gravity at a place and a value known as the *average normal gravity* at the same place. The idea underlying average normal gravity is this; gravity is imagined to be equal around the equator and likewise equal around each parallel of latitude. In other words average normal gravity is imagined to vary with latitude. Well, since actual gravity is not nearly so consistent, average normal gravity values will differ from the measured gravity at a place. The calculation of geoid heights using gravity anomalies, Δg, is usually done with the formula derived in 1849 by George Stokes, but its effectiveness depends on the accuracy of the modeling of a geoid.

There have been major improvements in the mapping of the geoid on both national and global scales over the past decades. NGS's geoid modeling has become more and more refined. These models supply values for N, a geoid height, for the official vertical datum at any place as long as its latitude and longitude are available. The latest of these produced by the NGS is GEOID12B. It supports conversion between NAD83 ellipsoidal heights and orthometric heights in NAVD88, *Guam Vertical Datum of 2004, GUVD04; Northern Marianas Vertical Datum of 2003, NMVD03; Puerto Rico Vertical Datum of 2002, PRVD02; American Samoa Vertical Datum of 2002, ASVD0, Virgin Islands Vertical Datum of 2009, VIVD09.* It will also support conversions for a new planned Hawaii vertical datum. Here are some of GEOID12B's predecessors.

The GEOID90 model was rolled out at the end of 1990. It was built using over a million gravity observations. The GEOID93 model, released at the beginning of 1993 utilized many more gravity values. Both provided users with geoid heights in a grid of 3 min of latitude by 3 min of longitude. The next model, GEOID96, had a grid that was 2 min of latitude by 2 min of longitude in the conterminous United States and a 2 min by 4 min grid in Alaska. It also differed from its two immediate predecessors in that it relied and was adjusted to measurements on benchmarks known as *GPSBMs.*

A word about GPSBMs is appropriate as they played a substantial role in the creation of GEOID96 and all the NGS geoid models since. About half a million NGS benchmarks have orthometric heights, H, derived from geodetic leveling. About 80,000 NGS benchmarks also have GPS/GNSS derived NAD83 ellipsoidal heights, h. However, only about a third of the benchmarks with both orthometric and ellipsoidal height values are known as GPSBMs. At those GPSBMs geoid heights are established through the relationship N = H – h. Some of those GPSBM geoid heights have been used in the development of a geoid models over the past decades. There are concentrations of GPSBMs in 5 states. For example, Minnesota has more than 4,000 GPSBMs spaced around 2 km from one another. Their distribution is more uneven and sparse in the other states where the spacing can reach 100 km or more. Where there are many GPSBMs closely spaced together the resulting geoid model is more reliable than it is in areas where they are few and far between.

GEOID96 was constructed, so it had an absolute agreement of ±5.5 cm (1 sigma) with a selection of 2951 GPSBMs. Improvement continued and GEOID99 with a grid of 1 min of latitude by 1 min of longitude was computed using 2.6 million gravity measurements and had an absolute agreement of ±4.6 cm (1 sigma) relative to 6169 GPSBMs. GEOID99 was followed by GEOID03, GEOID06, GEOID09, GEOID12, GEOID12A, and GEOID12B. Each superseded all the previous models and each of them utilized more GPSBMs than their predecessors.

Obviously there are many legacy geoid models, and as techniques and knowledge improved the models changed. It follows that older geoid models do not reflect the newest gravity data. A geoid model should only be used in conjunction with the specific reference frame for which it was designed. They cannot be mixed and matched. For example, the N values from GEOID12B are only properly used to calculate NAVD88 orthometric heights from NAD 83 (2011) epoch 2010.00 ellipsoid heights. 25,211 GPSBMs were used in the modeling of GEOID12B in the conterminous United States.

Two Categories of Geoid Models

There are two categories of geoid models provided by NGS. One category is known as *composite or hybrid geoid models*. GEOID96 was NGS's first hybrid geoid model, and the subsequent models up to and including the current GEOID12B are hybrid models. Hybrid models are created and transformed, so users with GPS/GNSS measurement and the model can compute heights in the current official vertical datum, NAVD88 for the conterminous United States and Alaska currently. However, there is usually a small difference between an orthometric height determined by differential leveling on a benchmark and an orthometric height on the same benchmark that is calculated using a GPS/GNSS ellipsoid height with the N value from NGS's hybrid model. Here is an example. NGS benchmark M 393, PID KK1340, has a NAD 83 (2011) ellipsoidal height, h, of 5506.44 ft, 1678.366 m, derived from a GPS/GNSS measurement. It has a published official N value of −54.18 ft, −16.513 m derived from GEOID12B. Here is the orthometric height calculated using these values:

$$H = h - N$$

$$H = 5506.44 \text{ ft} - (-54.18 \text{ ft})$$

$$H = 5560.62 \text{ ft}$$

The calculation does not produce exactly the same NAVD88 orthometric height, H, derived from leveling. That is, 5560.79 ft, 1694.931 m. In other words,

$$5560.79 \text{ ft} \neq 5560.62 \text{ ft}$$

They differ by 0.17 ft, 5 cm.

In the conterminous United States, the difference between an orthometric height from leveling and the height from calculations varies from about −0.66 ft, −20 cm to +0.57 ft, +18 cm depending on the location. As Geoid 12B is a best-fit model, the standard deviation is about +/−0.06 ft, 2 cm. This is due, in part, to the random errors in the GPS/GNSS measurements, accumulation of errors in differential leveling, and the distribution of the selected GPSBMs used in the geoid modeling. One way of accommodating these small discrepancies is to acquire an ellipsoidal height from a GPS/GNSS measurement on a benchmark with known orthometric height and compute a vertical shift, ΔV, as follows:

$$H = h - N + \Delta V$$

The geoid models in the second NGS category are known as *gravimetric geoid models*. The previously mentioned GEOID90 and GEOID93 were early gravimetric geoid models. Since then the creation of each of the hybrid geoid models from GEOID96 through GEOID12B started with an underlying gravimetric geoid model. The gravimetric models are then modified to fit GPS/GNSS ellipsoid heights on leveled benchmarks. The foundation of GEOID96 was the *G96SSS* gravimetric model. The foundation of GEOID12B was the *United States Gravimetric Geoid, USGG2012*, which covers the United States and its territories.

The hybrid and the gravimetric models differ. USGG2012 is defined entirely by gravity. GEOID12B cannot be, it is constrained to NAVD88 and the errors accumulated in its realization. One aspect of that constraint is the zero point at Father Point/Rimouski, which was, please recall, chosen to provide agreement with the International Great Lakes Datum of 1985 and to lessen map recompilation efforts.

Now that GRACE satellite data is available for comparison it turns out that NAVD88 zero surface is on average ~1.6 ft, ~50 cm, above the best gravimetric geoid as shown in Figure 3.10. Therefore, it follows that the difference between an orthometric height derived from the most up-to-date gravimetric geoid and a NAVD 88 orthometric height is ~1.6 ft, ~50 cm.

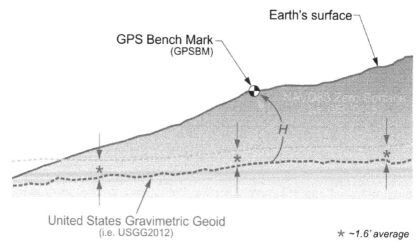

FIGURE 3.10
The difference between GEOID12B and USGG2012.

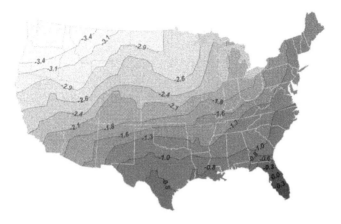

FIGURE 3.11
GPS/GNSS ellipsoidal heights—GRACE geoid heights—NAVD88 orthometric heights.

The comparison also reveals that the NAVD88 zero surface is tilted across the conterminous United States. The values in Figure 3.11 are calculated by finding the difference between GRACE-derived geoid heights, NAVD88 orthometric heights, and GPS/GNSS derived ellipsoidal heights at positions (Roman et al. 2010).

$$\text{Residual} = h - N - H$$

$$\text{Residual} = \text{GPS/GNSS ellipsoidal height}$$

$$- \text{GRACE geoid height} - \text{NAVD88 orthometric height}$$

(Roman et al. 2010, p. 2)

If the three values were correlated the results of the calculations would be zero. They are not. There are residuals, shown in Figure 3.11, which illustrate the extent of the trend of the tilt between the GRACE geoid and the NAVD88 zero elevation reference surface. The difference varies from +0.52 ft, +16 cm in Florida to −3.6 ft, −1.12 m in the Pacific Northwest. The same process reveals a tilt of up to ~6.6 ft, ~2 m across Alaska. Please note that this the situation even after the GRACE geoids have been transformed so their centers are coincident with the NAD83 center. If this had not been done, the shifts shown in Figure 3.11 would be substantially larger.

Please recall that without such a transformation the centers of the hybrid and the gravimetric models differ. The hybrid GEOID12B is the geoid reference for NAD83, as mentioned in Chapter 2 the GRS80 reference ellipsoid used for NAD83 is not geocentric. The gravimetric USGG2012 geoid

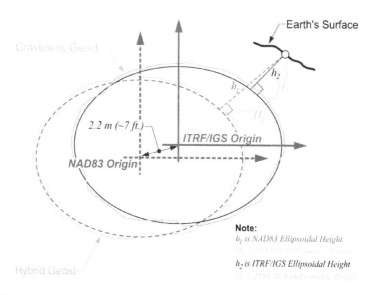

FIGURE 3.12
Gravimetric geoid and hybrid geoid orientations.

refers to the ITRF/IGS. ITRF/IGS uses the GRS80 ellipsoid too. However, in those reference frames it is oriented to the geocenter. A comparison between the NAD83 and the ITRF/IGS centers revealed a difference of approximately ~7 ft, ~2.2 m. As shown in Figure 3.12 that shift contributes to the difference in the heights from the two frames.

It is definitely not appropriate to use USGG2012 to convert between the NAD83 ellipsoid heights and NAVD88 orthometric heights. On the other hand, GEOID12B is designed for that purpose exactly.

Replacement of NAVD88

Technological advances, GPS/GNSS, GRACE, GOCE, and others have systematically improved measurement since NAVD88 was rolled out many years ago. The subsequent measurements have revealed inadequacies in the official vertical datum, which have led to plans to replace NAVD88 with a new geopotential datum by 2022. The new datum will be called the *North American-Pacific Geopotential Datum of 2022 (NAPGD2022)*. It is called a "geopotential" rather than a "vertical" datum because it includes information other than just orthometric heights, such as deflection of the vertical, surface gravity, and geoid heights (Dennis 2017). The NGS plan is known as *GRAV-D*,

Gravity for the Redefinition of the American Vertical Datum. One objective of the program is to build an unconstrained gravimetric geoid model that will support both orthometric and dynamic heights. However, the geoid will not only be modeled it will also be monitored. The masses of the planet are in a constant state of flux and thereby gravity changes. As gravity changes, heights change. NAVD88, a datum based on a static gravity model cannot account for that.

The realization of NAVD88 included conventional level work and the establishment of orthometric heights on hundreds of thousands of passive benchmarks. The benchmarks were frequently set along railroads and highways because they offered flat continuous pathways for differential levels. This approach makes the distribution of benchmarks uneven. It also makes them vulnerable. When tracks and lanes are reconstructed or improved thousands are destroyed or disturbed. It happens every year. Even when they survive regional distortions are inevitable, their heights can be invalidated without notice by crustal movement such as isostatic rebound, freezing and thawing, and uplift and subsidence. Subsidence is particularly troublesome in low-lying coastal areas along the Gulf of Mexico, Chesapeake Bay, and California agricultural regions. Some of NAVD88's original benchmark's heights have been updated or superseded with local releveling or GPS/GNSS. However, there has been no nationwide effort to readjust NAVD88 since its implementation and NGS does not have sufficient resources to check and/or replace existing passive benchmarks. This is particularly significant because users currently rely on those benchmarks for access to the datum.

The gravimetric geoid will be free to be the geopotential surface that best fits Mean Sea Level (MSL) in a least squares sense. As sea level rises NGS will likely update the geoid model, so that it fits the changing MSL. Users will notice orthometric heights changing as the vertical distance separating a given location and MSL decreases. This will be of particular importance in coastal areas.

References

Dennis, Michael. 2017. Email message to author, March 14.

Rappleye, H. S. *Manual of Geodetic Leveling, Special Publication No. 239*. Washington, DC: U.S. Department of Commerce, Coast and Geodetic Survey, 1948.

Roman, D. R., Y. M. Wang, J. S., and X. Li., Geodesy, geoids, & vertical datums: A perspective from the U.S. National Geodetic Survey. *Paper presented at FIG Congress 2010, Facing the Challenges—Building the Capacity*, Sydney, Australia, April 11–16, 2010.

Exercises

1. Which of the following can be depended upon to define the flow of water?
   a. From a higher ellipsoidal height to a lower ellipsoidal height
   b. From a higher geodetic height to a lower geodetic height
   c. From a higher orthometric height to a lower orthometric height
   d. From a higher dynamic height to a lower dynamic height

Explanation:

An interesting point, water would not necessarily flow from a higher ellipsoidal height to a lower ellipsoidal height. You need to use an equipotential surface for that. It would definitely flow downhill from a higher dynamic height to a lower dynamic height.

Answer is (d)

2. Which of the following letters is typically used to represent a height determined by a GPS/GNSS measurement that is available without reference to a geoid?
   a. H
   b. h
   c. N
   d. None of the above

Explanation:

Where N represents a geoid height between a geoid and the ellipsoid, it is a distance perpendicular to the ellipsoid to a geoid. If a geoid is above the ellipsoid, N is positive and if a geoid is below the ellipsoid, N is negative. H represents the orthometric height, which is the positive, upward distance from a geoid to the point on the Earth's surface measured along a plumb line. The sum of these two yields the ellipsoid height, also known as the geodetic height, h, that is the distance from the reference ellipsoid to the point on the surface of the Earth. If the ellipsoid's surface is below the surface of the Earth at the point the ellipsoidal height has a positive sign, if the ellipsoid's surface is above the surface of the Earth, at the point the ellipsoidal height has a negative sign. An ellipsoidal height is the height determined by a GPS/GNSS measurement that is available without reference to a geoid. The letter typically used to represent it is h.

Answer is (b)

3. Until the 1940s the Coast and Geodetic Survey stamped the elevation of a point on its monuments. Which of the following is one reason those elevations are now obsolete?
   a. They would be expressed in feet, not in meters
   b. The elevations would be based on the Krasovsky reference ellipsoid
   c. The elevations would not be in the North American Datum of 1988
   d. The elevations would be dynamic rather than orthometric

Explanation:

The elevations stamped on passive benchmark monuments before the establishment of NAVD88 are now superseded. The monument itself has not moved, but its official elevation has changed.

Answer is (c)

4. Which of the following heights is not measured along a line that is perpendicular to the ellipsoid?
   a. An ellipsoidal height
   b. A geoid height
   c. A geodetic height
   d. An orthometric height

Explanation:

A geodetic height is another name for an ellipsoidal height. Both the ellipsoidal, or geodetic height, and a geoid height are measured along lines perpendicular to the reference ellipsoid. The orthometric height, on the other hand, is not. It is measured along a plumb line, which is perpendicular to all of the equipotential surfaces through which it passes from a geoid to the point on the Earth's surface.

Answer is (d)

5. Which of these values would not have been affected by gravity?
   a. An ellipsoidal height of 2729.463 m
   b. 983.124 gals from a gravimeter measurement
   c. The measurement of hourly heights of the sea at a primary-control tide station
   d. An orthometric elevation of 5,176.00 ft derived from spirit leveling

Explanation:

An ellipsoidal height does not have a relationship with gravity, as does an orthometric elevation derived from spirit leveling. The orthometric elevation is measured from a geoid, which is an equipotential surface defined by gravity, and the leveling instrument itself is oriented to gravity when it is properly set up. The variation in tides is certainly affected by the gravity of the Earth, the Moon, and the Sun, as are the measurements of a gravimeter, which are of the acceleration of gravity itself.

Answer is (a)

6. Which of the following events contributed to a substantial increase in the utilization of spirit leveling over trigonometric leveling in precise geodetic surveying in the United States?

    a. The beginning of triangulation to cross the continent along the 39th parallel.

    b. The deterioration of the dock at Pointe-au-Père and the subsequent moving of the IGLD zero point to Rimouski, Quebec.

    c. The adjustment that established the Sea Level Datum of 1929.

    d. The establishment of GPS/GNSS as a fully operational system.

Explanation:

Congress mandated that the Coast Survey was to begin a transcontinental arc of triangulation to connect the surveys on the Atlantic coast with those on the Pacific coast in 1871. Trigonometric leveling was not sufficient to support the project, more vertical accuracy was needed. So, the *Coast Survey* under Benjamin Peirce depended on spirit leveling. Until that time their work had been restricted to the coasts. The work began at the Chesapeake Bay in 1878.

Answer is (a)

7. Which of the following was the first composite or hybrid geoid model produced by NGS?

    a. GEOID90

    b. GEOID93

    c. GEOID96

    d. GEOID03

Explanation:

There are two categories of geoid models provided by NGS. One category is known as *composite or hybrid geoid models*. GEOID96 was NGS's first hybrid geoid model, and the subsequent models up to and including the current GEOID12B are hybrid models. Hybrid models are created and biased, so users with GPS/GNSS measurement and the model can compute heights in the current official vertical datum, NAVD88 for the conterminous United States currently. GEOID90 and GEOID93 were early gravimetric geoid models.

Answer is (c)

8. When the heights of benchmarks from a spirit level circuit are compared with the orthometric heights of the same benchmarks calculated from GPS/GNSS observations they may differ even though they both are based on NAVD88. Which of the following could not contribute to the difference?

   a. The undulation of the GRS80 ellipsoid.

   b. Lack of balance between the foresights and the backsights when the spirit level circuit was run.

   c. Error in a calculated geoid height.

   d. The lack of parallelism in equipotential surfaces.

Explanation:

The accuracy of level work depends on the techniques and the care used. Methods such as balancing the forward and backsights, using one piece rods and so on, can improve results markedly, and not using these methods can lead to systematic error creeping into the work.

Also, equipotential surfaces are not parallel due to variation in the density of the Earth's crust, the Earth's rotation, and other effects. The convergence of equipotential surfaces prevents leveling from providing the differences between points as it is defined in orthometric heights. And the amount of the effect depends somewhat on the direction of the level circuit. The GRS80, unlike a geoid, does not undulate. It is a smooth mathematical surface.

Answer is (a)

9. Which of the following statements is correct?

   a. In the conterminous United States the GRS80 ellipsoid is always below the NAVD88 zero surface.

b. The heights of the Atlantic Ocean and the Pacific Ocean are the same.

c. Mean Sea Level and a geoid that best fits Mean Sea Level from a least squares point of view nevertheless deviates from Mean Sea Level up to 2 m at some places.

d. A geoid that best fits Mean Sea Level from a least squares point of view is always below the GRS80 ellipsoid all around the world.

Explanation:

The separation between surface the of a geoid that best fits Mean Sea Level from a least squares point of view and the smooth GRS80 ellipsoid worldwide varies from about +85 m west of Ireland to about −106 m, the latter in the area south of India near Ceylon. That geoid is sometimes above and sometimes below the GRS80 ellipsoid. In the conterminous United States, sometimes-abbreviated CONUS, the distances between that geoid and the ellipsoid are less. They vary from about −8 m to about −53 m. Geoid heights are usually symbolized, N. If a geoid is above the ellipsoid, N is positive if a geoid is below the ellipsoid, N is negative. It is negative because a geoid that best fits Mean Sea Level from a least squares point of view is actually underneath the ellipsoid throughout the conterminous United States. Around the world Mean Sea Level deviates up to 1, even 2, m from that equipotential surface.

Answer is (c)

10. What is a gal?
    a. A center fleeing force.
    b. An equipotential surface of the Earth's gravity field.
    c. A unit used to measure speed.
    d. A unit of acceleration equal to 1 cm per second per second.

Explanation:

The acceleration of gravity, that is the rate at which a falling object changes its velocity, is usually quantified in *gals*, a unit of measurement named for Galileo, who pioneered the modern understanding of gravity. What is a gal? Well, imagine an object traveling at 1 cm per second at the end of one second. If that object's speed increased to 2 cm per second by the end of the next second it would have accelerated 1 gal. Said another way, 1 gal is an acceleration of 1 cm per second per second.

Answer is (d)

4

Two Coordinate Systems

State Plane Coordinates

State Plane Coordinates rely on an imaginary flat reference surface with Cartesian axes. They describe measured positions by ordered pairs, expressed in northings, N, and eastings, E, or x- and y-coordinates. By the way it is important to keep in mind that $(x, y) = (E, N)$.

Despite the fact that the assumption of a flat Earth is fundamentally wrong, calculation of areas, angles, and lengths using latitude and longitude can be complicated, so plane coordinates persist. Therefore, the projection of points from the Earth's surface onto a reference ellipsoid and finally onto flat surfaces carries on.

In fact, many agencies of government, particularly those that administer state, county, and municipal databases prefer coordinates in their particular *State Plane Coordinate System, SPCS*. The systems are, as the name implies, state specific. In many states the system is officially sanctioned by legislation. Generally speaking, such legislation allows surveyors to use State Plane Coordinates to legally describe property corners. It is convenient. A Cartesian coordinate and the name of the officially sanctioned system are sufficient to uniquely describe a position. The same fundamental benefit makes the SPCS attractive to government; it allows agencies to assign unique coordinates based on a common, consistent system throughout its jurisdiction.

Map Projection

State Plane Coordinate Systems are built on *map projections*. Map projection means representing a portion of the actual Earth on a plane. Done for hundreds of years to create paper maps, it continues, but map projection today is most often a mathematical procedure done in a computer. However, even in an electronic world it cannot be done without distortion.

The problem is often illustrated by trying to flatten part of an orange peel. The orange peel stands in for the surface of the Earth. A small part, say a square a quarter of an inch on the side, can be pushed flat without much noticeable deformation. But when the portion gets larger problems appear.

Suppose a third of the orange peel is involved, as the center is pushed down the edges tear and stretch, or both. And if the peel gets even bigger, the tearing gets more severe. So, if a map is drawn on the orange before it is peeled, the map gets distorted in unpredictable ways when it is flattened. And it is difficult to relate a point on one torn piece with a point on another in any meaningful way.

These are the problems that a map projection needs to solve to be useful. The first problem is the surface of an ellipsoid, like the orange peel, is *non-developable.* In other words, flattening it inevitably leads to distortion that is very difficult to model consistently. So, a useful map projection ought to start with a surface that is *developable,* a surface that may be flattened without all that unpredictable deformation. It happens that a paper cone or cylinder both illustrate this idea nicely. They are illustrations only; models for thinking about the issues are involved.

If a right circular cone is cut from bottom to top up one of its elements that is perpendicular from the base, the cone can then be made completely flat without trouble. The same may be said of a cylinder cut up a perpendicular from base to base (Figure 4.1).

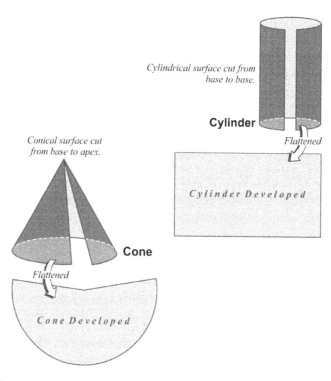

FIGURE 4.1
The development of a cylinder and a cone.

One could use the simplest case, a surface that is already developed, and a flat piece of paper. If the center of a flat plane is brought tangent to the Earth, a portion of the planet can be mapped on it. In other words, a portion of the Earth can be projected directly onto the flat plane. In fact this is the typical method for establishing an independent *local coordinate system*. These simple Cartesian systems are convenient and satisfy the needs of small projects. The method of projection, onto a simple flat plane, is based on the idea that a small section of the Earth, as with a small section of the orange mentioned previously, conforms so nearly to a plane that distortion on such a system is negligible.

Subsequently, local tangent planes have been long used by land surveyors. Such systems demand little if any manipulation of the field observations and the approach has merit as long as the extent of the work is small. But the larger the plane grows the more untenable it becomes. As the area being mapped grows the reduction of survey observations becomes more complicated as it must take account of the actual shape of the Earth. This usually involves the ellipsoid, a geoid and geographical coordinates, latitude and longitude. At that point, surveyors and engineers rely on map projections to mitigate the situation and limit the now troublesome distortion. However, a well-designed map projection can offer the convenience of working in plane Cartesian coordinates and still keep distortion at manageable levels.

The design of such a projection must accommodate some awkward facts. For example, while it would be possible to imagine mapping a considerable portion of the Earth using a large number of small individual planes, like facets of a gem, it is seldom done because when these planes are brought together they cannot be edge-matched accurately. They cannot be joined properly along their borders. The problem is unavoidable because the planes, tangent at their centers, inevitably depart more and more from the reference ellipsoid at their edges. As the distance between the ellipsoidal surface and the surface of the map on which it is represented grows, so does the distortion on the resulting flat map. This is true of all methods of map projection. Therefore, one is faced with the daunting task of joining together a mosaic of individual maps along their edges where the accuracy of the representation is at its worst. And even if one could overcome the problem by making the distortion the same on two adjoining maps, another difficulty would remain. Typically each of these planes has a unique coordinate system like the Local Geodetic Horizon System, LGHS or LGH illustrated in Chapter 3. The orientation of the axes, the scale, and the rotation of each one of these individual local systems will not be the same as those elements of its neighbor's coordinate system. Consequently there are both gaps and overlaps between adjacent maps and their attendant coordinate systems. Without a common reference system the difficulties of moving from map to map are compounded (Figure 4.2).

So the idea of a self-consistent local map projection based on small flat planes tangent to the Earth, or the reference ellipsoid, is convenient, but only

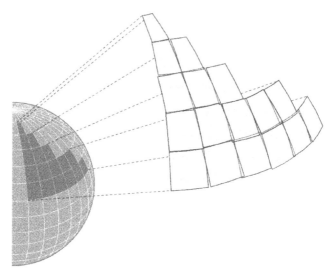

FIGURE 4.2
Local coordinate systems do not edge-match.

for small projects that have no need to be related to adjoining work. And as long as there is no need to venture outside the bounds of a particular local system it can be entirely adequate. But, generally speaking, if a significant area is involved in the work another strategy is needed. However, that is not to say that tangent plane map projections have no larger use. Please consider the tangent plane map projections that are used to map the polar areas of the Earth.

Polar Map Projections

These maps are generated on a large tangent plane touching the globe at a single point, the pole. Parallels of latitude are shown as concentric circles. Meridians of longitude are straight lines from the pole to the edge of the map. The scale is correct at the center. But just as the smaller local systems mentioned earlier, the farther you get from the center of the map the more they are distorted. These maps and this whole category of map projections are called *azimuthal*. The polar aspect of two of them will be briefly mentioned—the stereographic and the gnomonic. One clear difference in their application is the position of the imaginary *light source*.

A point light source is a useful device in imagining the projection of features from the Earth onto a developable surface. The rays from this light source can be imagined to move through a translucent figure representing the Earth and thereby project the image of the area to be mapped onto the mapping surface, similar to the projection of the image from film onto a screen. This is, of course, another model for thinking about map projection, an illustration.

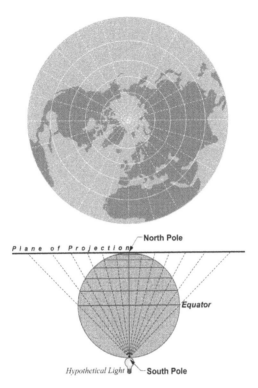

FIGURE 4.3
A stereographic projection, polar aspect.

In the case of the stereographic map projection, this point light source is exactly opposite the point of tangency of the mapping surface. In Figure 4.3 the North Pole is the point of tangency. The light source is at the South Pole.

On this projection, shapes are correctly shown. In other words, a rectangular shape can be expected to appear as a rectangular shape on the map with its right angles preserved. Map projections that have this property are said to be *conformal*.

In another azimuthal projection, the *gnomonic*, the point light source moves from opposite the tangent point to the center of the globe. The term *gnomonic* is derived from the similarity between the arrangement of meridians on its polar projection and the hour marks on a sundial. The gnomon of a sundial is the structure that marks the hours by casting its shadow on those marks. This projection is not conformal.

In Figures 4.3 and 4.4 the point at the center of the map, the tangent point, is sometimes known as the *standard* point. In map projection places where the map and the figure representing the Earth touch are known as standard lines or points. These are the only places on the map where the scale is exact. Therefore, *standard points* and *standard lines* are the only

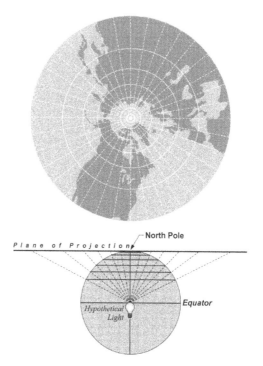

FIGURE 4.4
A gnomonic projection, polar aspect.

places on a map and the resulting coordinates systems derived from them are really completely free of distortion.

As mentioned earlier, a map projection's purpose informs its design. For example, the small individual plane projections first mentioned conveniently serve work of limited scope. Such a small-scale projection is easy to construct and can support Cartesian coordinates tailored to a single independent project with minimal calculations.

Plane polar map projections are known as azimuthal projections because the direction of any line drawn from the central tangent point on the map to any other point correctly represents the actual direction of that line. And the gnomonic projection can provide the benefit that the shortest distance between any two points on the ellipsoid, a *great circle*, can be represented on a gnomonic map as a straight line. It is also true that all straight lines drawn from one point to another on a gnomonic map represent the shortest distance between those points. These are significant advantages to navigation on air, land, and sea.

Decreasing that distortion is a constant and elusive goal in map projection. It can be done in several ways. Most involve reducing the distance between the map projection surface and the ellipsoidal surface. One way this is done to move the mapping surface from tangency with the ellipsoid and make it actually cut through it. This strategy produces a *secant* projection. A secant

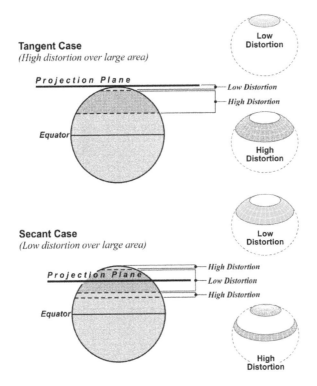

FIGURE 4.5
Minimizing distortion with a secant projection.

projection is one way to shrink the distance between the map projection surface and the ellipsoid. Thereby the area where distortion is in an acceptable range on the map can be effectively increased as shown in Figure 4.5.

Another strategy can be added to this idea of a secant map projection plane. This is the strategy used to produce the stereographic secant plane projection illustrated in the lower part of Figure 4.5. It is known as the *Universal Polar Stereographic (UPS)* system used to augment the *Universal Transverse Mercator* (UTM) projection, UPS is an application for which a flat plane projection is well suited. However, the area covered is large, so the distortion of UPS that varies from −6000 ppm at the poles to −3270 ppm at the edge of the map of the northern polar region and +1608 ppm at the edge of the map of the southern polar region makes it unsuitable for many purposes. So, to reduce the distortion even more, one can use one of those developable surfaces mentioned earlier, a cone or a cylinder along with a secant projection.

Both cones and cylinders have an advantage over a flat map projection plane. They are curved in one direction and can be designed to follow the curvature of the area to be mapped in that direction. Also, if a large portion of the ellipsoid is to be mapped, several cones or several cylinders may be used together in the same system to further limit distortion as shown in Figure 4.6.

FIGURE 4.6
Several cones establishing Lambert Conformal zones.

In that case, each cone or cylinder defines a *zone* in a larger coverage. This is the approach used in State Plane Coordinate Systems.

When a conic or a cylindrical map projection surface is made secant, it intersects the ellipsoid, and the map is brought close to its surface. The conic and cylindrical projections shown in Figure 4.7 cut through

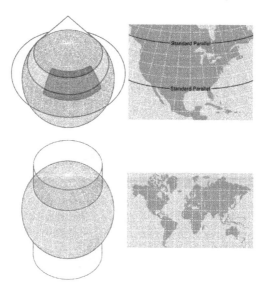

FIGURE 4.7
Secant projections with developable surfaces.

the ellipsoid. The map is projected both inward and outward onto it. And two *lines of exact scale*, standard lines, are created along the *small circles* where the cone and the cylinder intersect the ellipsoid. They are called small circles because they do not describe a plane that goes through the center of the Earth. Where the ellipsoid and the map projection surface touch, in this case intersect, there is no distortion. However, between the standard lines the map is under the ellipsoid and outside of them the map is above it. That means, between the standard lines a distance from one point to another is actually longer on the ellipsoid than it is shown on the map, and outside the standard lines a distance on the ellipsoid is shorter than it is on the map. Any length that is measured along a standard line is the same on the ellipsoid and on the map, which is why another name for standard parallels is *lines of exact scale*.

Choices

Here and in most mapping literature the cone and cylinder, the hypothetical light source, and other abstractions are mentioned because they are convenient models for thinking about the steps involved in building a map projection. Ultimately, the goal is very straightforward, relating each position on one surface, the reference ellipsoid, to a corresponding position on another surface as faithfully as possible and then flattening that second surface to accommodate Cartesian coordinates. In fact, the whole procedure is in the service of moving from geographic to Cartesian coordinates and back again. These days the complexities of the mathematics are handled with computers. Of course, that was not always the case. In 1932, two engineers in North Carolina's highway department, O.B. Bester and George F. Syme, appealed to the then *United States Coast and Geodetic Survey* (USC&GS, now NGS) for help. They had found that the stretching and compression inevitable in the representation of the curved Earth on a plane was so severe over long route surveys that they could not check into the USC&GS geodetic control stations across a state within reasonable limits. The engineers suggested that a plane coordinate grid system be developed that was mathematically related to the reference ellipsoid, but could be utilized using plane trigonometry.

Dr. Oscar Adams of the Division of Geodesy, assisted by Charles Claire, designed the first *State Plane Coordinate System* to mediate the problem. It was based on a map projection called the *Lambert Conformal Conic Projection*. Dr. Adams realized that it was possible to use this map projection and allow one of the four elements of area, shape, scale, or direction to remain virtually unchanged from its actual value on the Earth, but not all four. On a perfect map projection all distances, directions, and areas could be conserved. They would be the same on the ellipsoid and on the map. Unfortunately, it is not possible to satisfy all of these specifications simultaneously, at least not completely. There are inevitable choices. It must be decided which characteristic will be shown most correctly, but it will be done at the expense of the others. And there is no universal best decision. Still a solution that gives the most satisfactory results for a particular mapping problem is always available.

Dr. Adams chose the *Lambert Conformal Conic Projection* for the North Carolina system. On the Lambert Conformal Conic Projection parallels of latitude are arcs of concentric circles and meridians of longitude that are equally spaced straight radial lines, and the meridians and parallels intersect at right angles. The axis of the cone is imagined to be a prolongation of the polar axis. The parallels are not equally spaced because the scale varies as you move north and south along a meridian of longitude. Dr. Adams decided to use this map projection in which shape is preserved based on a developable cone.

Map projections in which shape is preserved are known as conformal or *orthomorphic*. Orthomorphic means right shape. In a conformal projection the angles between intersecting lines and curves retain their original form on the map. In other words, between short lines, meaning lines under about 10 miles, a 45° angle on the ellipsoid is a 45° angle on the map. It also means that the scale is the same in all directions from a point; in fact, it is this characteristic that preserves the angles. These aspects were certainly a boon for the North Carolina Highway engineers and benefits that all State Plane Coordinate users have enjoyed since. On long lines, angles on the ellipsoid are not exactly the same on the map projection. Nevertheless, the change is small and systematic. It can be calculated.

Actually, all three of the projections that were used in the designs of the original State Plane Coordinate Systems were conformal. Each system was based on the North American Datum 1927, NAD27 originally. Along with the *Oblique Mercator* projection, which was used on the panhandle of Alaska, the two primary projections were the Lambert Conic Conformal Projection and the *Transverse Mercator Projection,* also known as the *Gauss–Krüger Projection.* For North Carolina, and other states that are longest east–west, the Lambert Conic projection works best. State Plane Coordinate Systems in states that are longest north–south were built on the Transverse Mercator projection. There are exceptions to this general rule. For example, California uses the Lambert Conic projection even though the state could be covered more efficiently with Transverse Mercator. The Lambert Conic projection is a bit simpler to use, which may account for the choice.

The Transverse Mercator projection is based on a cylindrical mapping surface much like that illustrated in Figure 4.7. However, the axis of the cylinder is rotated, so that it is perpendicular with the polar axis of the ellipsoid as shown in Figure 4.8. Unlike the Lambert Conic projection the Transverse Mercator represents meridians of longitude as curves rather than straight lines on the developed grid. The Transverse Mercator projection is not the same thing as the UTM system. UTM was originally a military system that covers the entire Earth, and differs significantly from the Transverse Mercator system used in State Plane Coordinates.

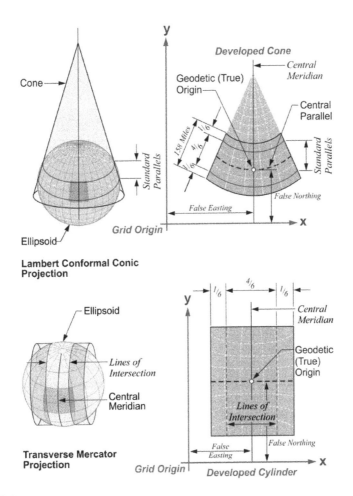

FIGURE 4.8
Two SPCS map projections.

The architect of both the Transverse Mercator projection, built on work by Gerardus Mercator, and the conformal conic projection that bears his name was Johann Heinrich Lambert, an eighteenth century Alsation mathematician. His works in geometry, optics, perspective, and comets are less known than his investigation of the irrationality of π. Surveyors, mappers, and cartographers know Lambert's mapping projections above all. It is especially remarkable that the projections he originated are used in every state of the United States, and both were first presented in his contribution, *Beiträge zum Gebrauche der Mathematik und deren Anwendung*, in 1772. Still the Lambert Conic projection was little used until the United States Coast and Geodetic Survey, encouraged by Dr. Adams and Charles Deetz, began publishing his theory and tables from which it could be applied in 1918.

Dr. Adams used these projections as the foundation of the State Plane Coordinate Systems (SPCS). SPCS are projected coordinate systems. That means that they include a projection type, a reference frame, and a linear unit. In choosing the projection type Dr. Adams wanted to have the advantage of conformality.

He also wanted to cover each state with as few zones as possible. A zone in this context is a belt across the state that has one Cartesian coordinate grid with one origin and is projected onto one mapping surface. One strategy that played a significant role in achieving that end was Dr. Adams's use of *secant* projections in both the Lambert and Transverse Mercator systems.

For example, using a single secant cone in the Lambert projection and limiting the extent of a zone, or belt, across a state to about 158 miles, approximately 254 km, Dr. Adams limited the distortion of the length of lines. Angles were preserved on the final product. For example, meridians of longitude and parallels of latitude were perpendicular to one another. There were minimal differences between the length of a measured line on the Earth's surface and the length of the same line on the map projection, minimal for the measurement technology of the day. In other words, the scale of the distortion was pretty small.

As illustrated in Figure 4.8 he placed $^4/_6$ th of the map projection plane between the standard lines, $^1/_6$ th outside at each extremity. The distortion was held to 1 part in 10,000. A maximum distortion in the lengths of lines of 1 part in 10,000. The difference between the length of a two-mile line on the ellipsoid and its representation on the map would be about 1 foot at the most, and the scale distortion would be the same in every direction from a point.

State Plane Coordinates were created to be the basis of a method that approximates geodetic accuracy more closely than the then commonly used methods of plane surveying. Today, surveying methods can easily achieve accuracies of 1 part in 100,000 and better, but the State Plane Coordinate Systems were designed in a time of generally lower accuracy and efficiency in surveying measurement. Today computers easily handle the lengthy and complicated mathematics of geodesy. But the first State Plane Coordinate System was created when such computation required sharp pencils and logarithmic tables. In fact, the original SPCS was so successful in North Carolina that similar systems were devised for all the states in the Union within a year or so. The system was successful because, among other things, it overcame some of the limitations of mapping on a horizontal plane while avoiding the imposition of strict geodetic methods and calculations. It managed to keep the distortion of the scale ratio under 1 part in 10,000 and preserved conformality. It did not disturb the familiar system of ordered pairs of Cartesian coordinates and it covered each state with as few zones as possible whose boundaries were constructed to follow portions of county lines as much as possible, with some exceptions like Alaska where the Transverse Mercator zones are four-degrees wide. The idea was that those relying on State Plane Coordinates could often work in one zone throughout a jurisdiction.

State Plane Coordinate Systems 27 to State Plane Coordinate Systems 83

In Figure 4.9 the current boundaries of the 123 State Plane Coordinates System zones are shown. In several instances they differ from the original zones. The boundaries shown in the figure are for *SPCS83*, the State Plane Coordinate System based on NAD83 and its reference ellipsoid GRS80. The foundation of the original State Plane Coordinate System, *SPCS27* was NAD27 and its reference ellipsoid Clarke 1866. As mentioned in earlier chapters NAD27 geographical coordinates, latitudes and longitudes, differ significantly from those in NAD83. In fact, conversion from geographic coordinates, latitude and longitude, to grid coordinates, x and y and back is one of the three fundamental conversions in the State Plane Coordinate System. It is important because the whole objective of the SPCS is to allow the user to work in plane coordinates, but still have the option of expressing any of

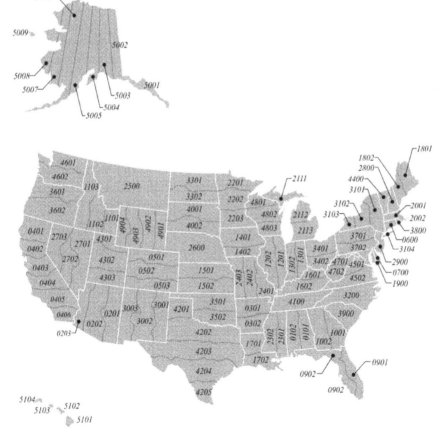

FIGURE 4.9
SPCS83 zones.

the points under consideration in either latitude and longitude or State Plane Coordinates without significant loss of accuracy. Therefore, when geodetic control was migrated from NAD27 to NAD83 the State Plane Coordinate System had to go along.

When the migration was undertaken in the 1970s it presented an opportunity for an overhaul of the system. Many options were considered but in the end just a few changes were made. One of the reasons for the conservative approach was the fact that 37 states had passed legislation supporting the use of State Plane Coordinates. All the states, except Alaska and Florida, continue to use the same map projection type for all zones. Nevertheless some zones got new numbers and some of the zones changed. The zones in Figure 4.9 are numbered in the SPCS83 system known as *FIPS*. FIPS stands for Federal Information Processing Standard, and each SPCS83 zone has been given a FIPS number. These days the zones are often known as *FIPS zones*. SPCS27 zones did not have these FIPS numbers. As mentioned earlier the original goal was to keep each zone small enough to ensure that the scale distortion was 1 part in 10,000 or less. However, when the SPCS83 was designed that scale was not maintained in some states like Montana.

In five states some SPCS27 zones were eliminated altogether and the areas they had covered consolidated into one zone or added to the adjoining zones. In three of those states the result was one single large zone. Those states are South Carolina, Montana, and Nebraska. In SPCS27 South Carolina and Nebraska had two zones, in SPCS83 they have just one, FIPS zone 3900 and FIPS zone 2600 respectively. Montana previously had three zones. It now has one, FIPS zone 2500. Therefore, because the area covered by these single zones has become so large they are not limited by the 1 part in 10,000 standard, the scale along the central parallel has become 1 part in 1646. California eliminated zone 7 and added that area to FIPS zone 0405, formerly zone 5. Two zones previously covered Puerto Rico and the Virgin Islands. They now have one. It is FIPS zone 5200. In Michigan, three Transverse Mercator zones were entirely eliminated.

In both the Transverse Mercator and the Lambert projection the positions of the axes are similar in all SPCS zones. As you can see in Figure 4.8 each zone has a central meridian. These central meridians are true meridians of longitude near the geometric center of the zone. In Transverse Mercator the central meridian is also the *projection axis*, the line along which the scale is at its minimum in relation to the reference ellipsoid. The central meridian of the Lambert Conformal Conic is not the projection axis, where the scale is the least. That is the *central parallel* as illustrated in Figure 4.8

Please note that the central meridian is not the y-axis. If it were the y-axis, negative coordinates would result. To avoid them the actual y-axis is moved far to the west of the zone itself. In the old SPCS27 arrangement the y-axis was 2,000,000 ft west from the central meridian in the Lambert Conic projection and 500,000 ft in the Transverse Mercator projection. In the SPCS83 design, those constants have been changed. The most common values are

600,000 m for the Lambert Conic and 200,000 m for the Transverse Mercator. However, there is a good deal of variation in these numbers from state-to-state and zone-to-zone. In all cases the y-axis is still far to the west of the zone and the x-axis, also known as the baseline, is far to the south of the zone. Where the x-axis and y-axis intersect is the grid origin of the zone and that is always south and west of the zone itself. This configuration of the axes ensures that all State Plane Coordinates occur in the first quadrant, so there are no negative State Plane Coordinates. The replacement of NAD83 circa 2022 will not be likely to alter the current SPCS83 zone extents unless a state wishes to revise them. However, the false northings and false eastings values may again be changed to provide a difference in the coordinate values significant enough to prevent confusion with the superseded SPCS83 coordinates.

There are five parameters necessary to fully define a State Plane Coordinate zone. Two parameters that must be defined are the latitude and longitude of the *geodetic (aka true) origin*. In the Transverse Mercator projection the geodetic origin longitude always places it on the projection axis, but its latitude is arbitrary and is chosen for convenience. Likewise the geodetic origin of a Lambert Conformal Conic map projection is typically on the central meridian, but its latitude need not place it on the central parallel.

The other necessary parameters include the scale at the projection axis: the false northing and the false easting.

There is sometimes even further detail in the name of particular State Plane Coordinates. As refinements are made to NAD83 the new adjustments are added as a suffix to the SPCS83 label. For example, SPCS83/99 would refer to State Plane Coordinates that were based on a revision to NAD83 from 1999.

It is important to note that the fundamental unit for SPCS27 is the U.S. survey foot, and for SPCS it is the meter. The conversion from meters to the U.S. survey feet is correctly accomplished by multiplying the measurement in meters by the fraction 3937/1200.

In the following sections the most typical conversions used in State Plane Coordinates will be addressed:

1. Conversion from geodetic lengths to grid lengths.
2. Conversion from geographic coordinates, latitude and longitude, to grid coordinates.
3. Conversion from geodetic azimuths to grid azimuths.
4. Conversion from SPCS to ground coordinates.

Geodetic Lengths to Grid Lengths

This brings us to the scale factor, also known as the *K factor* and the *projection factor*. It was this factor that the original design of the State Plane Coordinate system sought to limit to 1 part in 10,000. As implied by that effort, scale factors are ratios that can be used as multipliers to convert ellipsoidal lengths, also known as *geodetic distances*, to lengths on the map projection surface, also known as *grid distances*, and vice versa. In other words, the geodetic length of a line, on the ellipsoid, multiplied by the appropriate scale factor will give you the grid length of that line on the map. And the grid length multiplied by the inverse of that same scale factor would bring you back to the geodetic length again.

While referring to Figure 4.8 it is interesting to note that on the projection used on most states that are longest from east to west, that is the Lambert Conic, the scale factor for east–west lines is constant. In other words, the scale factor is the same all along the line. One way to think about this is to recall that the distance between the ellipsoid and the map projection surface never changes east to west in that projection. On the other hand along a north–south line the scale factor is constantly changing on the Lambert Conic. And it is no surprise then to see that the distance between the ellipsoid and the map projection surface is always changing north to south line in that projection. But looking at the Transverse Mercator projection, the projection used most on states longest north to south, the situation is exactly reversed. In that case, the scale factor is the same all along a north–south line and changes constantly along an east–west line.

Both the Transverse Mercator and the Lambert Conic used a secant projection surface and originally restricted the width to 158 miles. These were two strategies used to limit scale factors when the State Plane Coordinate Systems were designed. Where that was not optimum, the width was sometimes made smaller, which means the distortion was lessened. As the belt of the ellipsoid projected onto the map narrows, the distortion is reduced. For example, Connecticut is less than 80 miles wide north to south. It has only one zone. Along its northern and southern boundaries, outside of the standard parallels, the scale factor is 1 part in 40,000, a four-fold improvement over 1 part in 10,000. And in the middle of the state the scale factor is 1 part in 79,000, nearly an eight-fold improvement. On the other hand, the scale factor was allowed to get a little bit smaller than 1 part in 10,000 in Texas. By doing that the state was covered completely with five zones. And among the guiding principles in 1933 was covering the states with as few zones as possible and having zone boundaries follow county lines. Still it requires ten zones and all three projections to cover Alaska.

In Figure 4.10 a typical 158-mile State Plane Coordinate zone is represented by a grid plane of projection cutting through the ellipsoid of reference. As mentioned earlier between the intersections of the standard lines, the grid is under the ellipsoid. There, a distance from one point to another is longer on the ellipsoid than on the grid. This means that right in the middle of a SPCS zone the scale factor is at its minimum, this is along the central parallel in the Lambert Conic. In the middle a typical minimum SPCS scale factor is not less than 0.9999, though there are exceptions. Here the distance from one

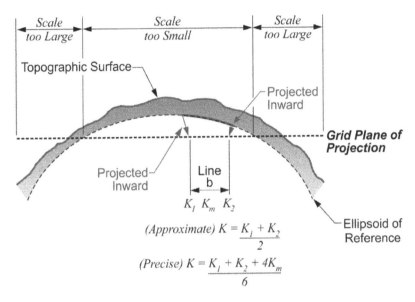

FIGURE 4.10
Scale factor.

point to another is longer on the ellipsoid than it is on the grid. Outside of the intersections the grid is above the ellipsoid where a distance from one point to another is shorter on the ellipsoid than it is on the grid. There at the edge of the zone a maximum typical SPCS scale factor is generally not more than 1.0001.

When SPCS27 was current, scale factors were interpolated from hard-copy tables published for each state. In the tables for states in which the Lambert Conic projection was used scale factors change north–south with the changes in latitude. In the tables for states in which the Transverse Mercator projection was used scale factors change east–west with the changes in the x-coordinate. Today scale factors are not interpolated from tables for SPCS83. For both the Transverse Mercator and the Lambert Conic projections they are calculated directly from equations.

To illustrate the use of these factors please consider line b in Figure 4.10 to have a length on the ellipsoid of 130,210.44 ft, a bit over 24 miles. That would be its geodetic distance. The official unit of SPCS83 is the meter, however, feet are often used. Suppose that a scale factor for that line was 0.9999536, then the grid distance along line b would be

$$Geodetic\ Distance * Scale\ Factor = Grid\ Distance$$

$$130,210.44\ ft * 0.999953617 = 130,204.40\ ft$$

The difference between the longer geodetic distance and the shorter grid distance here is a little more than 6 ft. That is actually better than 1 part in 20,000; please recall that the 1 part in 10,000 ratio was considered the maximum. Distortion lessens and the scale factor approaches 1 as a line nears a standard parallel.

In the Lambert projection an east–west line, a line that follows a parallel of latitude, has the same scale factor at both ends and throughout. However, a line that bears in any other direction will have a different scale factor at each end. A north–south line will have a great difference in the scale factor at its north end compared with the scale factor of its south end. In this vein, please note the approximate formula near the bottom of Figure 4.11:

$$K = \frac{K_1 + K_2}{2}$$

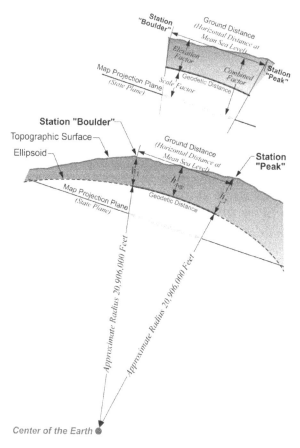

FIGURE 4.11
SPCS combined (grid) factor.

where:

  K is the scale factor for a line

  K_1 is the scale factor at one end of the line

  K_2 is the scale factor at the other end of the line

Scale factor varies with the latitude in the Lambert projection. For example, suppose the point at the north end of the 24-mile line is called Stormy and has a geographic coordinate of

$$37°\,46'\,00.7225"\,N \text{ latitude}$$

$$103°\,46'\,35.3195"\,W \text{ longitude}$$

and at the south end the point is known as Seven with a geographic coordinate of

$$37°\,30'\,43.5867"\,N \text{ latitude}$$

$$104°\,05'\,26.5420"\,W \text{ longitude}$$

The scale factor for point Seven 0.99996113 and the scale factor for point Stormy is 0.99994609. It happens that point Seven is further south and closer to the standard parallel than is point Stormy, and it therefore follows that the scale factor at Seven is closer to 1. It would be exactly 1 if it were on the standard parallel, which is why the standard parallels are called *lines of exact scale*. The typical scale factor for the line is the average of the scale factors at the two end points:

$$K = \frac{K_1 + K_2}{2}$$

$$0.99995361 = \frac{0.99996113 + 0.99994609}{2}$$

Deriving the scale factor at each end and averaging them is the usual method for calculating the scale factor of a line. The average of the two is sometimes called K_m. In Figure 4.10 there is another formula for calculating a more precise scale factor by using a weighted average. In this method, K_1 is given a weight of 1, K_m a weight of 4, and K_2 a weight of 1. No matter which method is used the result is still called the scale factor.

But that is not the whole story when it comes to reducing distance to the State Plane Coordinate grid. Measurement of lines must always be done on the topographic surface of the Earth, and not on the ellipsoid. Therefore the first step in deriving a grid distance must be moving a measured line from the Earth to the ellipsoid. In other words, converting a distance measured on the topographic surface to a geodetic distance on the reference ellipsoid. This is done with another ratio that is also used as a multiplier. Originally, this factor had a rather unfortunate name. It used to be known as the sea level factor in SPCS27. It was given that name because as you may recall from Chapters 2 and 3 when NAD27 was established using the Clarke 1866 reference ellipsoid the distance between the ellipsoid and the geoid was declared to be zero at Meades Ranch in Kansas. That meant that in the middle of the country the sea level surface, the geoid, and the ellipsoid were coincident by definition. And since the Clarke 1866 ellipsoid fit the United States quite well the separation between the two surfaces, the ellipsoid and geoid only grew to about 12 m anywhere in the country. With such a small distance between them many practitioners at the time took the point of view that, for all practical purposes, the ellipsoid and the geoid were in the same place. And that place was called sea level. Hence reducing a distance measure on the surface of the Earth to the ellipsoid was said to be reducing it to sea level.

Today that idea and that name for the factor are misleading because, of course, the GRS80 ellipsoid on which NAD83 is based is certainly not the same as Mean Sea Level. Now the separation between the geoid and ellipsoid can grow as large as −53 m. And technology by which lines are measured has improved dramatically. Therefore, in SPCS83 the factor for reducing a measured distance to the ellipsoid is known as the *ellipsoid factor*. In any case, both the old and the new name can be covered under the name the *elevation factor* or *height factor*.

Regardless of the name applied to the factor, it is a ratio. The ratio is the relationship between an approximation of the Earth's radius, and that same approximation with the mean ellipsoidal height of the measured line added to it. For example, please consider station Boulder and station Peak illustrated in Figure 4.11.

Boulder

39°59′29.1299″N latitude

105°15′39.6758″W longitude

Peak

40°01′19.1582″N latitude

105°30′55.1283″W longitude

The distance between these two stations is 72,126.21 ft. This distance is sometimes called the *ground distance* or the *horizontal distance at mean elevation*. In other words it is not the slope distance but rather the distance between them corrected to an averaged horizontal plane, as is a common practice. For practical purposes this is then the distance between the two stations on the topographic surface of the Earth. On the way to finding the grid distance Boulder to Peak there is the interim step, calculating the geodetic distance between them, that is the distance on the ellipsoid. We need the elevation or height factor and here is how it is determined.

The ellipsoidal height of Boulder, h_1, is 5437 ft. The ellipsoidal height of Peak, h_2, is 9099 ft. The approximate radius of the Earth, traditionally used in this work, is 20,906,000 ft. The elevation factor is calculated as

$$Elevation\ Factor = \frac{R}{R + h_{avg}}$$

$$Elevation\ Factor = \frac{20,906,000\ ft}{20,906,000\ ft + 7268\ ft}$$

$$Elevation\ Factor = \frac{20,906,000\ ft}{20,913,268\ ft}$$

$$Elevation\ Factor = 0.99965247$$

This factor then is the ratio used to move the ground distance down to the ellipsoid, down to the geodetic distance.

$$Ground\ Distance\ Boulder\ to\ Peak = 72,126.21\ ft$$

$$Geodetic\ Distance = Ground\ Distance * Elevation\ Factor$$

$$Geodetic\ Distance = 72,126.21 * 0.99965247$$

$$Geodetic\ Distance = 72,101.14\ ft$$

It is possible to refine the calculation of the elevation factor by using an average of the actual radial distances from the center of the ellipsoid to the end points of the line, rather than the approximate 20,906,000 ft. In the area of stations Boulder and Peak the average ellipsoidal radius is actually a bit longer, but within the continental United States such variation will not cause a calculated geodetic distance to differ significantly. However, it is worthwhile to take care to use the ellipsoidal heights of the stations when calculating the elevation factor, rather than the orthometric heights.

In calculating the elevation factor in SPCS27 no real distinction is made between ellipsoid height and orthometric height. However, in SPCS83 the averages of the ellipsoidal heights at each end of the line can be used for h_{avg}. If an ellipsoid height is not directly available it can be calculated from the formula

$$h = H + N$$

where:
  h is the ellipsoid height
  H is the orthometric height
  N is the geoid height

As mentioned previously converting a geodetic distance to a grid distance is done with an averaged scale factor:

$$K = \frac{K_1 + K_2}{2}$$

In this instance the scale factor at Boulder is 0.99996703 and at Peak it is 0.99996477.

$$0.99996590 = \frac{0.99996703 + 0.99996477}{2}$$

Using the scale factor it is possible to reduce the geodetic distance 72,101.14 ft to a grid distance:

$$Geodetic\ Distance * Scale\ Factor = Grid\ Distance$$

$$72,101.14\ ft * 0.99996590 = 72,098.68\ ft$$

There are two steps: first from ground distance to geodetic distance using the elevation factor and second from geodetic distance to grid distance using the scale factor. These two steps can be combined into one. Multiplying the elevation factor and the scale factor produces a single ratio that is usually known as the *combined factor* or the *grid factor*. Using this grid factor the measured line is converted from a ground distance to a grid distance in one jump. Here is how it works. In the example above the elevation factor for the line from Boulder to Peak is 0.99965247 and the scale factor is 0.99996590:

$$Grid\ Factor = Scale\ Factor * Elevation\ Factor$$

$$0.99961838 = 0.99996590 * 0.99965247$$

Then using the grid factor the ground distance is converted to a grid distance

$$Grid\ Distance = Grid\ Factor * Ground\ Distance$$

$$72,098.68\ ft = 0.99961838 * 72,126.21\ ft$$

Also the grid factor can be used to go the other way. If the grid distance is divided by the grid factor it is converted to a ground distance.

$$Ground\,Distance = \frac{Grid\,Distance}{Grid\,Factor}$$

$$72,126.21\,ft = \frac{72,098.68\,ft}{0.99961838}$$

Geographic Coordinates to Grid Coordinates

Please consider again two previously mentioned stations: Stormy and Seven. Stormy has an NAD83 geographic coordinate of

Latitude 37° 46′ 00.7225″ N latitude

Longitude 103° 46′ 35.3195″ W longitude

and Seven has an NAD83 geographic coordinate of

Latitude 37° 30′ 43.5867″ N latitude

Longitude 104° 05′ 26.5420″ W longitude

Finding the State Plane Coordinates of these stations can be accomplished online. The NGS website https://www.ngs.noaa.gov/TOOLS provides several programs, and among them is one named State Plane Coordinates. It has two convenient routines that allow the user to convert a station from SPCS to latitude and longitude and vice versa in both NAD27 and NAD83 and discover its scale factor and convergence angle. It is interesting to note that the site also includes free routines for conversion from SPCS27 to SPCS83. The conversion is actually done in a three-step process. The SPCS27 coordinate is converted to an NAD27 geographic coordinate, a latitude and longitude. Next the NAD27 geographic coordinate is transformed into an NAD83 geographic coordinate, and finally the NAD83 geographic coordinate becomes a SPCS83 coordinate. This procedure is common to nearly all GIS software. In any case using the State Plane Coordinates software it is possible to find the SPCS83 coordinates for these two stations:

Stormy:	N 428,305.869
	E 1,066,244.212
Seven:	N 399,570.490
	E 1,038,989.570

Both these are in meters, the native unit of SPCS83. The original SPCS27 design was based on the use of the U.S. Survey foot as its unit of measurement.

That remains the appropriate unit for that system today. However, SPCS83 is a bit more complicated in that regard. Although the fundamental unit for SPCS83 is the meter, when it comes to converting coordinates from meters to feet, one of the two conversions is called for. One is the conversion from meters to the U.S. Survey feet. The other conversion is from meters to International feet. The International foot, so named because an international agreement was established to define one inch equal to 2.54 cm exactly, is equal to 0.3048 m. And that version of the foot was adopted across the United States, except at the United States Coast and Geodetic Survey. In 1959 instead of forcing the USC&GS to refigure and republish all the control station coordinates across the country it was given a reprieve under the Federal Register Notice 24 FR 5348. The agency was allowed to retain the old definition of the foot, which is the U.S. Survey foot, in which 1 m is 39.37 in. exactly. Another way of saying it is a U.S. Survey foot is 1200/3937 m.

It was decided that the USC&GS could continue to use the U.S. Survey foot until the national control network was readjusted, but following the adjustment the agency was to switch over to the International foot. Things did not quite work out that way. When NAD83 was fully established the agency that had replaced the USC&GS, the NGS, mandated that the official unit of all the published coordinate values would be the meter. Then in 1986, NGS declared that it would augment its publication of State Plane Coordinates in meters with coordinates for the same stations in feet. Which foot? The version legislated by the state in which the station was found.

In practical terms this means that in states such as the 11 that have chosen the U.S. Survey foot and the 6 that have chosen the International foot it is clear which should be used. However, 14 states do not specify the version of the foot that is official for their SPCS. And the remaining 19 states have no legislation on the State Plane Coordinates at all.

The difference between the two foot-units may seem an academic distinction, but please consider the conversion of station Stormy. It happens that the station is in the Colorado Zone 0503, or South Zone, and that state has chosen the U.S. Survey feet. It follows that the correct coordinate values for the station in the U.S. Survey feet are

Stormy: N 1,405,200.17
                     E 3,498,169.55

However, if the metric coordinates for Stormy were mistakenly converted to International feet, they would be N 1,405,202.98 and E 3,498,176.55. The distance between the correct coordinate and the incorrect coordinate is more than 7½ ft with the largest difference occurring in the easting.

Conversion from Geodetic Azimuths to Grid Azimuths

As was mentioned in Chapter 1 meridians of longitude converge and that fact has an effect on the directions of lines in SPCS. To illustrate the rate of that convergence, please consider the east and west sides of a one-square mile figure somewhere in the middle of the conterminous United States. Suppose that the two sides were both meridians of longitude on the surface of the Earth and the direction of both lines was the same, that is, geodetic north. However, when that square mile figure is projected onto a Lambert Conic or Transverse Mercator SPCS their directions would no longer be equal. Their azimuths would suddenly differ by about 1 m of arc. And unless one of the sides happens to follow the central meridian of the zone, neither of their azimuths would be grid north.

In SPCS the direction known as grid north is always parallel to the central meridian for the zone. The east and west lines of the square mile in the example follow meridians on the surface of the Earth. Meridians converge at the pole; they are not parallel to one another. In SPCS, north is grid north and the lines of the grid are parallel to each other. They must also be parallel to one another and the central meridian of the zone, so clearly geodetic north and grid north are not the same. In fact, it is only on that central meridian that grid north and geodetic north coincide. Everywhere else in the zone they diverge from one another and there is an angular distance between them. In SPCS27, that angular distance was symbolized with the Greek letter theta, θ, in the Lambert Conic projection and with delta alpha, $\Delta\alpha$, in the Transverse Mercator projection. However, in SPCS83 convergence is symbolized with gamma, γ, in both the Lambert Conic and the Transverse Mercator projections.

In both map projections, east of the central meridian grid north is east of geodetic north and the convergence angle is positive. West of the central meridian grid north is west of geodetic north and the convergence angle is negative.

The angle between geodetic and grid north, the convergence angle, grows as the point gets further from the central meridian. It also gets larger as the latitude of the point increases. The formula for calculating the convergence in the Transverse Mercator projection is

$$\gamma = (\lambda_{cm} - \lambda) \sin \phi$$

where:
    λ_{cm} is the longitude of the central meridian
    λ is the longitude through the point
    ϕ is the latitude of the point

The formula for calculating the convergence angle in the Lambert Conic projection is very similar, it is

$$\gamma = (\lambda_{cm} - \lambda) \sin \phi_0$$

where:

λ_{cm} is the longitude of the central meridian
λ is the longitude through the point
ϕ_0 is the latitude of the center of the zone

As an example, here is the calculation of the convergence angle for station Stormy. It is in the Colorado South Zone on a Lambert Conic projection where the longitude of the central meridian is longitude 105° 30′ 00″ W and ϕ_0 is latitude 37° 50′ 02.34″ N

$$\gamma = (\lambda_{cm} - \lambda) \sin \phi_0$$

$$\gamma = (105°\,30'00'' - 103°\,46'\,35.3195'') \sin 37°\,50'02.34''$$

$$\gamma = (01°\,43'24.68'') \sin 37°\,50'02.34''$$

$$\gamma = (1°\,43'24.68'')\, 0.6133756$$

$$\gamma = +1°\,03'25.8''$$

The angle is positive as expected east of the central meridian.

The formula used to convert a geodetic azimuth to a grid azimuth includes the convergence angle, and another element.

$$\text{grid azimuth} = \text{geodetic azimuth} - \text{convergence} + \text{the second term}$$

Another way of stating the same formula is

$$t = \alpha - \gamma + \delta$$

where:

t is the grid azimuth
α is the geodetic azimuth
γ is the convergence angle
δ is the second term

The second term is included because lines between stations on the ellipsoid are curved and that curvature is not completely eliminated when the geodetic azimuth line is projected onto the SPCS grid (Figure 4.12).

Several things affect the extent of the curve of the projected geodetic azimuth line on the grid. The direction of the line and the particular map projection from which the grid was created are two of those elements. For instance, a north–south line does not curve at all when projected onto the Lambert Conic grid and neither does an east–west line when projected onto

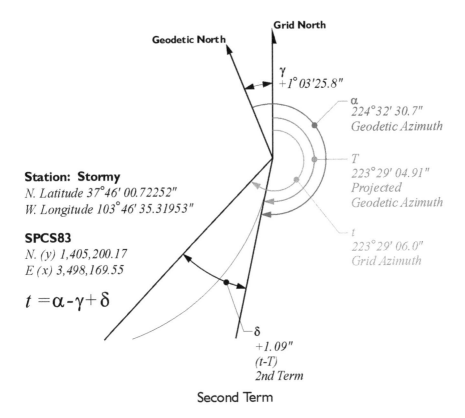

Grid North

Geodetic North

γ
$+1°03'25.8"$

α
$224°32'30.7"$
Geodetic Azimuth

T
$223°29'04.91"$
Projected
Geodetic Azimuth

t
$223°29'06.0"$
Grid Azimuth

Station: Stormy
N. Latitude 37°46' 00.72252"
W. Longitude 103° 46' 35.31953"

SPCS83
N. (y) 1,405,200.17
E (x) 3,498,169.55

$t = \alpha - \gamma + \delta$

δ
$+1.09"$
$(t-T)$
2nd Term

Second Term

FIGURE 4.12
Second term.

the Transverse Mercator grid. However, in both cases the more a line departs from these cardinal courses the more it will curve on the grid. In fact the maximum curvature in each projection occurs on lines that are parallel to their standard lines. That means that an east–west line would have the largest curve in a Lambert Conic projection and a north–south line would have the largest in a Transverse Mercator projection.

Another factor that affects the size of the curve is the distance of the line from the center of the zone. In a Transverse Mercator projection the farther a line is from the central meridian the more it will curve. In the Lambert Conic projection the farther a line is from the central parallel of latitude through the center of the zone, ϕ_0, the more it will curve. Finally, in both map projections the longer the line the more it will curve. It follows, therefore, that long lines at the boundaries between zones depart the most from straight lines on the grid. A two-mile north–south line in a Transverse Mercator SPCS will deviate about 1 arc second from a straight line. In the Lambert Conic SPCS an east–west line of that length will deviate about 1 arc second from a straight line.

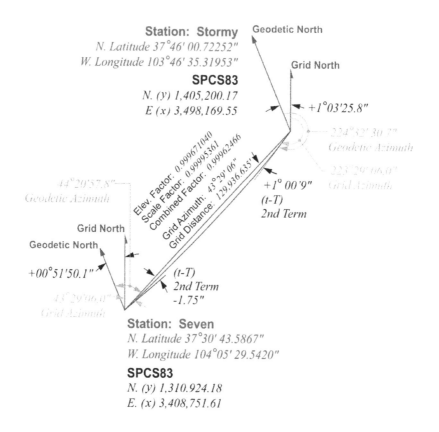

FIGURE 4.13
Stations Stormy and Seven.

Even though the distance between station Seven and station Stormy is approximately 24 miles, the departure from a chord between the two is only about 1″ at Stormy and 2″ at Seven. Please note, however, that the grid bearing of the line between the two is approximately N43 1/2°E rather than east–west (Figure 4.13).

This correction for curvature is sometimes known as *t-T*, the *arc to chord* correction and the *second term*. The correction is small and pertinent to the most precise work. However, it is important to note that like the convergence itself, the second term comes into play only when there is a need to convert a conventionally observed azimuth into a grid azimuth in SPCS. Where optical surveying data are used that will almost certainly be required. On the other hand, if the work is done with GPS/GNSS observations, or where field observations are not involved at all, convergence and the second term are not likely to affect the calculation of the work.

State Plane Coordinate Systems to Ground Coordinates

When a State Plane Coordinate is assigned to a station on the Earth the coordinate is not really on the station itself. As described above, the coordinate is on a grid. The grid is most often below and sometimes above the actual location of the point it intends to represent. Simply put the point is on the Earth and the coordinate is not. This rather inconvenient fact can be, and sometimes is, ignored. However, it is valuable to remember that while SPCS is generally designed to provide scale distortion no worse than 1 part in 10,000 or so, modern measurement is routinely better than that. It is therefore common for users of the system to find that their measurement from one place to another turns out to be longer, or shorter, than an inverse between the SPCS coordinates indicates. Such results sometimes lead to an assumption that the whole system is shot through with unacceptable inaccuracies. That is certainly not the case. It is also usual that once the actual cause of the difference between ground and grid dimensions is fully understood a convenient resolution is sought. Frequently, the resolution is bringing the State Plane Coordinates from the grid to the ground.

This is done by extending the idea mentioned at the end of the section on conversion from geodetic lengths to grid lengths. The idea is dividing the grid distance by the grid or combined factor to find the ground distance. This concept is sometimes applied to more than the single line where the particular grid factor is actually correct. It is used to convert many lines and many points over an entire project. In those instances a single grid factor calculated near the center of a project is used. This one grid factor is intended to convert grid distances between points to the ground distances ignoring the changes in the scale factor and the elevation factor from point to point.

The coordinates that result from this approach are not State Plane Coordinates, and are often truncated to avoid being confused with actual SPCS. The typical truncation is dropping the first two digits from the northings and eastings of these project coordinates. This strategy in some ways defeats the purpose of the SPCS, which when used correctly offers a reasonably accurate approximation of geodetic positions that is consistent over a large area. However, fixing one grid factor for a project returns a user to the sort of tangent plane system mentioned earlier. Such a project coordinate system cannot be joined along its edges with a neighboring system without difficulty. This difficulty cannot be avoided because the plane created near the elevation of the center of the project inevitably departs more and more from the reference ellipsoid at the edges. The advantages of a large secant map projection plane are removed. Therefore, as long as one stays within the now local system, the work can progress smoothly but outside of the area it will not match other systems.

It is interesting to note that some governmental organizations have established localized projections, often at the county level, to bring grid coordinates closer to the ground. The objective has been to provide a coordinate basis that is more convenient for building their GIS.

Common Problems with State Plane Coordinates

As mentioned earlier the official native format of SPCS83 coordinates is meters. However, reporting in feet is often required. Many states prefer U.S. Survey feet, that is, Nebraska, Wyoming, Colorado, California, Connecticut, Indiana, Maryland, North Carolina, and Texas. Other states specify International feet, that is, Arizona, Michigan, Montana, Oregon, South Carolina, and Utah. Still others have taken no official action on the issue. Nevertheless, clients in any state may request coordinates in either format.

If an error is suspected in converting SPCS83 coordinates from meters to feet, look to the easting of the coordinate. Since the false easting in SPCS is quite large, it is there that the discrepancy will be most obvious as was illustrated in the example conversion of station Stormy.

Another common problem stems from the periodic readjustments performed by NGS. As mentioned in Chapter 2 NAD83 has been subject to refinements since it replaced NAD27. These improvements are largely due to the increasing amount of GPS/GNSS information available and are denoted with a suffix, such as NAD83/91, the latter number referring to the year of the readjustment. Since SPCS83 is based on NAD83 these adjustments result in new State Plane Coordinates as well. It is therefore feasible that one county may use say NAD83/86 coordinates and an adjoining county may use NAD83/94 coordinates. The result may be a different coordinate assigned to the same station, both in NAD83, but differing as the result of each being in a different adjustment. The solution is to take care with the year of adjustment when using published coordinates.

When there is a discrepancy of millions of meters or feet between the eastings of coordinates of points that are certainly not hundreds of miles apart, the error may be attributable to SPCS27 coordinates among SPCS83 coordinates or vice versa. There were substantial changes made to the false easting when the datums were changed. Another possible culprit for the condition could be the coordinates of one SPCS83 zone being combined with coordinates from the adjoining zone.

Universal Transverse Mercator Coordinates

A plane coordinate system that is convenient for GIS work over large areas is the Universal Transverse Mercator, or UTM system. UTM with the Universal Polar Stereographic system covers the world in one consistent system. In terms of scale it is four times less precise than typical State Plane Coordinate Systems. Yet the ease of using UTM and its worldwide coverage makes it very attractive for work that would otherwise have to cross many different SPCS zones. For example, nearly all National Geospatial-Intelligence Agency

(NGA) (formerly *National Imagery and Mapping Agency, NIMA* and formerly the *Defense Mapping Agency*) topographic maps, USGS quad sheets, and many aeronautical charts show the UTM grid lines.

It is often said that UTM is a military system created by the United States Army. In fact, several nations, and the North Atlantic Treaty Organization, *NATO*, played roles in its creation after World War II. At that time the goal was to design a consistent coordinate system that could promote cooperation between the military organizations of several nations. Before the introduction of UTM, allies found that their differing systems hindered the synchronization of military operations.

Conferences were held on the subject from 1945 to 1951 with representatives from Belgium, Portugal, France, and Britain and the outlines of the present UTM system were developed. By 1951 the U.S. Army introduced a system that was very similar to that currently used.

The UTM projection divides the world into 60 zones. Actually one could say there are 120 zones since each of the 60 zones are divided into a Northern Hemisphere portion and Southern Hemisphere portion at the equator. The numbering of the zones begins at longitude 180°E, the International Date Line, and increases sequentially toward the east. Zone 1 is from 180°W longitude to 174°W longitude, Zone 2 is from 174°W longitude to the 168°W longitude, and so on. The conterminous United States are within UTM zones 10–19. Each zone has a central meridian in the middle. For example, in Zone 1 from 180°W longitude to the 174°W longitude the central meridian is 177°W longitude, so each zone extends 3° east and west from its central meridian. The central meridian for Zone 2 is 171°W longitude (Figure 4.14).

Here is a convenient way to find the zone number for a particular longitude. Consider west longitude negative and east longitude positive, add 180° and divide by 6. Any answer greater than an integer is rounded to the next highest integer and you have the zone. For example, Denver, Colorado is near 105°W. Longitude, −105°.

$$-105° + 180° = 75°$$

$$\frac{75°}{6} = 12.50$$

Round up to 13

Therefore, Denver, Colorado is in UTM zone 13.

All UTM zones have a width of 6° of longitude, which is ~319 miles, ~514 km, in the middle of the United States. From north to south the zones extend from 84°N latitude to 80°S latitude. Originally the northern limit was at 80°N latitude and the southern 80°S latitude. On the south the latitude is a small circle that conveniently traverses the ocean well south of Africa,

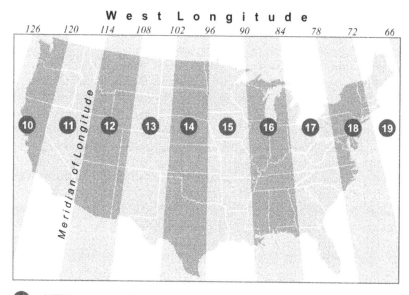

10 = UTM zone number

FIGURE 4.14
UTM zones in the conterminous United States.

Australia, and South America. However, 80°N latitude was found to exclude parts of Russia and Greenland and was extended to 84°N latitude.

These zones nearly cover the Earth, except the polar regions, which are covered by two azimuthal polar zones called the Universal Polar Stereographic, *UPS* projection. The foundation of the UTM zones is a secant Transverse Mercator projection very similar to those used in some State Plane Coordinate Systems. The UTM secant projection gives approximately 180 km between the lines of exact scale where the cylinder intersects the ellipsoid. The scale factor grows from 0.9996 along the central meridian of a UTM zone to 1.00000 at 180 km to the east and west. Please recall that SPCS zones are usually limited to about 158 miles and, therefore, have a smaller range of scale factors than do the UTM zones. In State Plane coordinates, the scale factor is usually no more than 1 part in 10,000. In UTM coordinates it can be as large as 1 part in 2,500.

There is no particular datum or reference ellipsoid stipulated for UTM coordinates. One can obtain 1983 UTM coordinates by referencing the UTM zone constants to the GRS80 ellipsoid of NAD83 in the United States. However, UTM is a global system and that approach is not applicable in other countries where the appropriate national datum was likely used.

As mentioned earlier, unlike any of the systems previously discussed every coordinate in a UTM zone occurs twice, once in the Northern Hemisphere and once in the Southern Hemisphere. This is a consequence of the fact that

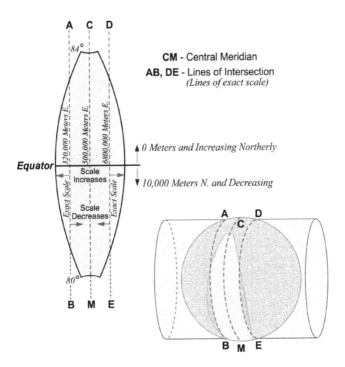

FIGURE 4.15
UTM zones around the world.

there are two origins in each UTM zone. The origin for the portion of the zone north of the equator is moved 500 km west of the intersection of the zone's central meridian and the equator. This arrangement ensures that all of the coordinates for that zone in the Northern Hemisphere will be positive. The origin for the coordinates in the Southern Hemisphere for the same zone is also 500 km west of the central meridian, but it is not at its intersection with the equator, it is 10,000 km south of it, close to the South Pole. This orientation of the origin guarantees that all of the coordinates in the Southern Hemisphere are in the first quadrant and are positive. In both hemispheres and for all zones the easting (the x-value) of the central meridian is 500,000 m at the central meridian as shown in Figure 4.15. The developed UTM grid is defined in meters.

In the official version of the UTM system there are actually more divisions in each UTM zone than the north–south demarcation at the equator. As shown in Figure 4.16 each zone is divided into 20 *latitudinal bands*. Each of the bands includes 8° of latitude and is lettered from C on the south to X on the north. Latitudinal band X is a bit longer than 8°; remember the extension of the system from 80°N latitude to 84°N latitude. That all went into band X. It is also interesting that the letters I and O are not included. They resemble one and zero too closely. In any case the letters and numbers

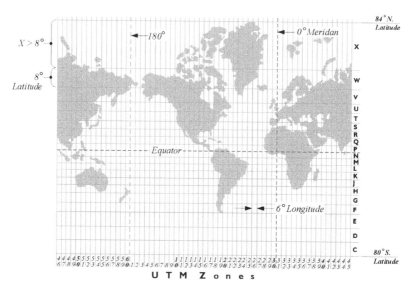

FIGURE 4.16
A UTM zone.

shown in Figure 4.16 are used in type of spatial address system known as the *Military Grid Reference System, MGRS* to identify a quadrangle in the UTM and UPS projections. A column of a quadrangle is identified by a zone number and a row by the letter. For example, The Department of the Interior building at 1849 C Street NW in Washington DC is in UTM zone 18 North and it is in latitudinal band S. So its alphanumeric string is 18S even though the UTM coordinate of the building in meters is 4307036 N and 322859 E in zone 18N. The *United States National Grid, USNG,* is quite similar to the MGRS but it is specifically referenced to NAD83, whereas the MGRS system is not.

A word or two about the polar zones that round out the UTM system. The Universal Polar Stereographic, *UPS* are conformal azimuthal stereographic projections like those mentioned earlier. The projection has two zones. The north zone covers the North Pole and the south zone the South Pole. The projection is on a secant plane at a pole and perpendicular to the minor axis of the reference ellipsoid. The projection lines originate from the opposite pole. As shown in Figure 4.17, the minimum scale factor is 0.994 at the pole. It increases to 1.0016076 at 80° latitude from each pole. The scale factor is constant along any parallel of latitude. The line of exact scale is at 81° 06′ 52.3″ N latitude at the North Pole and 81° 06′ 52.3″ S latitude at the South Pole. In both cases the pole is given a false easting and northing: the x-coordinate (easting) of the pole is 2,000,000 m and the y-coordinate (northing) of the pole is 2,000,000 m. The reference ellipsoid at both the North and South Poles is the International Ellipsoid.

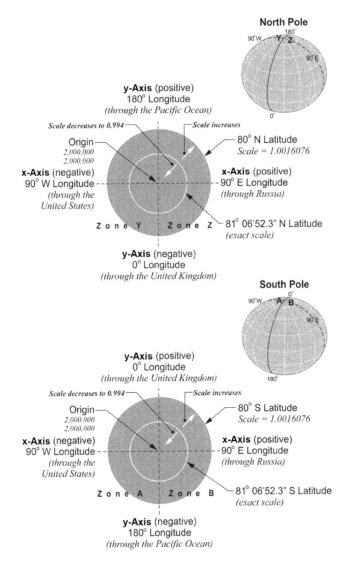

FIGURE 4.17
The Universal Polar Stereographic (UPS) projection.

At the North Pole and at the South Pole the x-axis lies along the meridians 90°W and 90°E. The easting values start at 2,000,000 at the pole and increase along the 90°E meridian. At the North Pole the y-axis lies along the meridians 0° and 180° and the northings start from 2,000,000 at the pole and increase along the 180° meridian. At the South Pole the y-axis also lies along the meridians 0° and 180° and the northings start from 2,000,000 at the pole but increase along the 0° meridian.

Exercises

1. Which of the following general statements about the State Plane
   Coordinate Systems in the United States is incorrect?

   a. There are three map projections used in SPCS in the United States.

   b. The map projections used in the SPCS cause meridians of longi-
      tude to appear as straight lines on the SPCS grid.

   c. Some states utilize more than one map projection in their SPCS.

   d. The axis of the Lambert cone is coincident with the Earth's polar
      axis; the axis of the Transverse Mercator cylinder is perpendicu-
      lar to it.

Explanation:

Every state, Puerto Rico, and the U.S. Virgin Islands has their own plane rectangular coordinate system. The State Plane Coordinate Systems are based on
the Transverse Mercator projection, an Oblique Mercator projection, and the
Lambert Conic map projection grid.

  Generally speaking, states with large east–west extent use the Lambert
Conic projection. This system uses a projection cone that intersects the ellipsoid at standard parallels. The axis of the cone is imagined to be a prolongation of the polar axis of the Earth. When the cone is developed, that is,
opened to make a plane, the ellipsoidal meridians become straight lines that
converge at the cone's apex. The apex is also the center of the circular lines
that represent the projections of the parallels of latitude.

  Some states use both the Lambert Conic and the Transverse Mercator
projections for individual zones within the state system. Some rely on the
Transverse Mercator projection alone. The Transverse Mercator projection
uses a projection cylinder whose axis is imagined to be parallel to the earth's
equator and perpendicular to its axis of rotation. It intersects the ellipsoid
along standard lines parallel to a central meridian. However, after the cylinder is developed all the projected meridians and parallels become curved
lines.

Answer is (b)

2. Which of the following statements does not properly describe a
   change in the State Plane Coordinate Systems from SPCS27 to
   SPCS83?

   a. There have been changes in the distances between the SPCS
      zones and their grid origins.

b. SPCS83 coordinates are expressed in more units than were SPCS27 coordinates.

c. There are fewer SPCS83 zones than there were SPCS27 zones in some states.

d. In SPCS83 the scale factor limit of 1 part in 10,000 has been exceeded for the first time.

Explanation:

In both SPCS27 and SPCS83, eastings are reckoned from a grid axis placed far west of the coordinate zone, so all coordinates remain positive. Northings are reckoned from a baseline far to the south of the zone for the same reason. Placing the y-axis 2,000,000 ft west from the central meridian in the Lambert Conic projection and 500,000 ft in the Transverse Mercator projection was usual in SPCS27. Those distances are now often 600,000 m for the Lambert Conic and 200,000 m for the Transverse Mercator in SPCS83. There is some variation in these SPCS83 distances, but there are no negative State Plane Coordinates in either SPCS27 or SPCS83 because the design of both assures that all coordinates fall in the first quadrant.

The native unit for SPCS27 coordinates is the U.S. Survey Feet. The native unit for SPCS83 coordinates is meters. SPCS83 coordinates are also expressed in either the U.S. Survey Feet or International Feet, depending on the state.

In five states some SPCS27 zones were eliminated altogether and the areas they had covered consolidated into one zone or added to adjoining zones. In three of those states the result was one single large zone. Those states are South Carolina, Montana, and Nebraska. The large single zones caused the 1 part in 10,000 scale factor standard to be exceeded in these states. However, it was not the first time. The standard had been exceeded by SPCS27 in several zones previously.

Answer is (d)

3. Which of the following statements correctly describes an aspect of conformal map projection in SPCS?

a. It ensures that the angle between two lines on the map projection surface is the same as it is on the reference ellipsoid regardless of their length.

b. In a conformal map projection the scale is the same in all directions from a point.

c. Convergence in a conformal map projection is always much less than 1°.

d. The Lambert Conic projection is a conformal projection but the Transverse Mercator projection is not.

Explanation:

Both the Lambert Conic and the Transverse Mercator projections used in SPCS are conformal. Conformality means that an angle between short lines on the ellipsoid is preserved after it is mapped onto mapping projection plane. This feature allows the shapes of small features to look the same on the map as they do on the earth. Map projections can be equal area, azimuthal, equidistant, or conformal. It would be great if one map projection could simultaneously possess all of these desirable characteristics at once, but that, unfortunately, is not possible. Conformality was built into the design of the State Plane Coordinate System from the beginning. It allows the transformation of points in such a way that, for all practical purposes, the angles between short lines on the projection surface are the same as they are on the reference ellipsoid because the scale is the same in all directions from a point. However, as the lengths of the lines grow a difference in the angles between them does become significant. This difference is an artifact of the curvature of lines on the ellipsoid and a correction known variously as the t-T, the arc to chord correction or the second term comes into play. Conformality does not limit the convergence of meridians, which is a function of the size of the SPCS zone.

Answer is (b)

4. Which of the following statements about the scale factor in SPCS is true?

a. The scale factor changes constantly along the central meridian of a Transverse Mercator SPCS zone.

b. Scale factor ratios are 1 part in 10,000 in SPCS zones.

c. A scale factor is used in SPCS to convert a length on the reference ellipsoid to a length on the map projection plane grid.

d. The scale factor does not change along the central meridian of a Lambert Conic SPCS zone.

Explanation:

The ratio of the length between two stations on the map projection plane and that same length on the reference ellipsoid is the scale factor. When the gap between the map projection plane and the reference ellipsoid does not change along the length of a line the scale factor does not change. When the gap between the two surfaces varies along the length of a line the scale factor varies. Therefore, on the conic Lambert projection the scale factor is constant along east–west lines and changes continuously along north–south lines. Since the central meridian in a conic projection is north–south, and the gap between the map projection plane and the reference ellipsoid varies along such a line, the scale factor varies. On the cylindrical Transverse Mercator projection the scale factor is constant along north–south lines and changes continuously along east–west lines. The central meridian in a cylindrical projection is north–south, and the gap between the map projection plane and the reference ellipsoid does not vary along such a line, the scale factor is constant.

It is not correct to say that the scale factor for SPCS zones has strictly adhered to the design criterion of 1 part in 10,000. For example, of those states that utilized the Lambert Conic projection in SPCS27 there were eight zones across the United States in which the scale factor ratio was smaller than 1 part in 10,000 along their central meridians. The situation has changed somewhat in SPCS83, for example, Montana, Nebraska, and South Carolina now have one zone each, instead of several with the attendant change in the scale factor. But while the objective limit was originally 1 part in 10,000, in fact, it has not been maintained universally as the SPCS system has evolved.

Answer is (c)

5. If a line on a secant map projection plane crossed one of the standard lines, what would be the difference between the scale factor at the end between the standard lines and the scale factor at the end outside the standard lines?

   a. The scale factor at the station outside the standard lines would be greater than 1; the scale factor for the station between the standard lines would be less than 1.

   b. For an east–west line that crossed a standard parallel in a Lambert Conic projection the scale would be the same at both ends of the line.

   c. For a north–south line that crossed a standard line of exact scale in a Transverse Mercator projection the scale would be the same at both ends of the line.

   d. The scale factor at the station outside the standard lines would be less than 1; the scale factor for the station between the standard lines would be greater than 1.

Explanation:

In a perfect map projection the scale factor would be 1 everywhere. Since that is not possible, in a secant projection the scale factor is less than 1 between standard lines and greater than 1 outside the standard lines. The only place where the scale factor is 1 in such a projection is along the standard lines themselves. The larger the variation in scale factors that is allowed in the original design of the system, the larger their range above and below 1.

Neither an east–west line on the Lambert Conic projection nor a north–south line on the Transverse Mercator projection would cross a standard line.

Answer is (a)

6. Which of the following statements about convergence angles in SPCS is not correct?

  a. As the distance from the central meridian increases, the absolute values of the convergence angles increase.

  b. As the length of a line connecting two stations increases, the absolute values of the convergence angles at each end station increase.

  c. As the direction of a line connecting two stations approaches cardinal, north–south or east–west, the absolute values of the convergence angles at each end station are not affected.

  d. As the latitude decreases, the absolute values of the convergence angles decrease.

Explanation:

In both the Transverse Mercator and Lambert Conic projections the absolute value of the convergence angle increases with the distance from the central meridian. It also increases with the latitude. In Lambert Conic zones the effect of latitude is constant for the zone, but it is greater in more northerly zones. In Transverse Mercator zones the effect of the latitude grows constantly from south to north. Since convergence angles are tied to the position of the stations on the grid, they are not affected by the length or the direction of the lines connecting those stations.

Answer is (b)

7. Which of the following statements about the second term, also known as t-T and arc-to-chord, correction in SPCS is not correct?

  a. As the distance from the central meridian in a Transverse Mercator projection and the distance from the central parallel of latitude, ϕ_0, in the Lambert Conic projection increases, the absolute value of the second term angles increase.

  b. As the length of a line connecting two stations increases, the absolute values of the second term angles at each end station increases.

c. As the direction of a line connecting two stations becomes more nearly parallel with a zone's standard lines, the absolute values of the second term angles at each end station increase.

d. As the latitude decreases, the absolute values of the second term angles decrease.

Explanation:

In both projections, Transverse Mercator and Lambert Conic the absolute values of the second term corrections at each end of a line are at their maximum when the line is at the edge of a zone and parallel with the standard lines. A north–south line does not exhibit any curve at all when projected onto the Lambert Conic grid and neither does an east–west line when projected onto the Transverse Mercator grid. However, in both cases the more a line departs from these cardinal courses the more it will curve on the grid. That means that an east–west line would have the largest curve in a Lambert Conic projection and a north–south line would have the largest in a Transverse Mercator projection. Again in a Transverse Mercator projection the farther a line is from the central meridian the more it will curve. In the Lambert Conic projection the farther a line is from the central parallel of latitude through the center of the zone, ϕ_0, the more it will curve. A longer line will have a larger second term correction at its ends than a shorter line.

The second term correction is usually small and ignored for much work. Its characteristics are essentially the same in SPCS27 and SPCS83. While there is a relationship between the position of a line on a particular SPCS zone and the second term correction, there is none between the latitude of the line and the size of the second term correction.

Answer is (d)

8. The combined factor is used in SPCS conversion. How is the combined factor calculated?

a. The scale factor is multiplied by the elevation, aka height, factor.

b. The scale factor is divided by the grid factor.

c. The grid factor added to the elevation, aka height, factor, and the sum is divided by two.

d. The scale factor is multiplied by the grid factor.

Explanation:

The grid factor changes with the ellipsoidal height of the line. It also changes with its location in relation to the standard lines of its SPCS zone. The grid factor is derived by multiplying the scale factor by the elevation factor. The product is nearly 1 and is known as either the grid factor or the combined factor. There is a different combined factor for every line in the correct application of SPCS.

Answer is (a)

9. Which value is closest to the typical change in the false easting between SPCS27 and SPCS83?

   a. The typical difference between the false easting in the Lambert Conic is about 6 miles and the typical difference in the Transverse Mercator is about 30 miles.

   b. The typical difference between the false easting in the Lambert Conic is about 1,400,000 m and the typical difference in the Transverse Mercator is about 3,000,000 m.

   c. The typical difference between the false easting in the Lambert Conic is about 9600 ft and the typical difference in the Transverse Mercator is about 4800 ft.

   d. The typical difference between the false easting in the Lambert Conic is about 47,600 ft and the typical difference in the Transverse Mercator is about 31,500 ft.

Explanation:

To avoid negative coordinates in SPCS the y-axis was moved west of the zone itself. In the SPCS27 the typical false easting from the y-axis to the central meridian was 2,000,000 ft in the Lambert Conic projection and 500,000 ft in the Transverse Mercator projection. There were substantial changes made to the false easting when the datums were changed. The most common values in SPCS83 are 600,000 m for the Lambert Conic, a shift of about 6 miles. The typical SPCS83 Transverse Mercator false easting is 200,000 m, a shift of about 30 miles. However, there is a good deal of variation in these numbers from state-to-state and zone-to-zone.

Answer is (a)

10. Which of the following does not correctly state a difference between the Transverse Mercator coordinate system used in SPCS and the Universal Transverse Mercator coordinate system?

    a. The UTM system of coordinates was designed and established by the United State Army in cooperation with NATO member nations. The Transverse Mercator SPCS was established by the United States Coast and Geodetic Survey.

b. The official unit of measurement in UTM system of coordinates is the meter. The official unit in Transverse Mercator SPCS83 is the International foot.

c. The UTM system of coordinates is augmented by UPS to complete its worldwide coverage. The Transverse Mercator SPCS does not have such a broad scope.

d. The scale factor in the UTM system of coordinates can reach 1 part in 2,500. In the Transverse Mercator SPCS the scale factor is usually no more than 1 part in 10,000.

Explanation:

Both the UTM system and the Transverse Mercator system in SPCS use meter as their official unit of measurement.

Answer is (b)

11. Which of the following statements correctly describe the line along which the scale is at its minimum in relation to the reference ellipsoid in the SPCS?

a. Along a line of exact scale in the Transverse Mercator

b. Along a standard parallel in the Lambert Conic

c. Along the projection axis in Transverse Mercator and Lambert Conic

d. Along the central meridian in the Lambert Conic projection

Explanation:

Scale is at its minimum in relation to the reference ellipsoid along the projection axis not along the lines of exact scale. In Transverse Mercator the central meridian is the projection axis. The central meridian of the Lambert Conformal Conic is not the projection axis, that is, the central parallel of latitude.

Answer is (c)

5

The Rectangular System

Some assert that the *Public Land Survey System, PLSS,* is a coordinate system. Others strongly maintain that it is certainly not. Suffice it to say that for more than 200 years it has been, and continues to be, a system to divide land and describe property across most of the United States. It was originally established by the *Land Ordinance Act of May 20, 1785.* The system has been incrementally improved through practical innovation and government regulation since. The Department of the Treasury had jurisdiction over the Public Land Survey System from its beginnings until 1812. It was a sensible arrangement since the government was poor in cash and rich in the land. The land came to be known as the *public domain,* meaning land that was once, or is even now, owned by the Federal Government of the United States. What was required was a practical, reliable, and unambiguous method of disposal. The fledgling Federal Government needed to do several things. It needed to pay soldiers, often with land instead of cash. It needed to disperse the population on the land to realize its potential. Put simply, it needed to raise money.

From the very beginning of the system it has always been the policy of the Federal Government that land in the public domain must be surveyed before it is sold to others, that is, before it is *patented.* Therefore, the United States embarked on the most extensive surveying project ever undertaken. The work began in Ohio in the autumn of 1785. On April 25, 1812 the Congress created the *General Land Office (GLO)* to survey and administer the PLSS. Initially the GLO did the work under a contract system. The contractors were Deputy Surveyors that worked for Surveyors General across the country. The surveying was not actually done by government employees until 1910. Later, the Bureau of Land Management was created in the Department of the Interior in 1946. It has been in charge of the system since.

The PLSS was and is an eminently practical attempt to cover a large portion of the curved surface of the Earth with a Cartesian grid with 1-mile squares, *sections.* The degree to which it has succeeded can be most clearly seen when flying over the Central and Western United States. The orderly patchwork quality of the landscape is a testament to those who created and those who maintained the Rectangular System.

Leaving aside the question of whether it is a coordinate system or not, the PLSS belongs to this book. It belongs here because of its prominent place in the definition of the lands and the GIS systems in the 30 states that were created from the public domain. These lands were acquired by purchase, treaty, or cession, and once amounted to more than 1.8 billion acres. The majority of that land has now been laid out into sections and passed from the Federal Government into other hands, but its measurement is still in most cases governed by the rules of the PLSS.

The Public Land Surveying System is a huge subject. The small portion of it that will be presented here is intended to offer just a basic understanding of a few elements of the topic that are pertinent to practitioners of GIS. Even though the application of the system can be complex, the design has always been straightforward and practical. It is that design, as currently practiced, that will be the emphasis here.

From the beginning the PLSS has been adaptable. It had to be. It needed to be able to apply to a variety of conditions on the land and be capable of changing with the times. It also needed to be as unambiguous as possible to prevent misunderstandings and boundary disputes. It has achieved these goals remarkably well.

The Initial Points

In a sense, *initial points* are the origins of the PLSS (Figure 5.1). They represent the intersection of two axes. There are 32 initial points in the conterminous United States and there are 5 in Alaska. Initial points were first mentioned in the written *Manuals of Instructions* in 1881. There were several of these manuals. The first official version was issued by the GLO in 1851. They were the vehicles by which the Commissioner of the GLO communicated the methods of survey to the Surveyors General. They reflect the evolution of the systems surveying procedures. However, by the time the establishment of initial points was mentioned in a manual many of the 37 in place today had already been set. It said,

> Initial points from which the lines of the public surveys are to be extended must be established whenever necessary under such special instructions as may be prescribed by the Commissioner of the General Land Office. The locus of such initial points must be selected with great care and due consideration for their prominence and easy identification, and must be established astronomically. (GLO 1881, p. 35)

In other words, they were set somewhat arbitrarily as needed and were assigned latitudes and longitudes derived astronomically.

East and west from the initial point a *Baseline* was laid out. This Baseline was intended to follow an astronomically determined parallel of latitude. And it did so as nearly as possible with corner monuments being established every half mile, which is every *40 chains*. The chain was, and is, the native unit of the PLSS system. A Gunter's chain is 66 ft long, and therefore, 40 chains is 2640 ft, or half a mile. It is also important to note that 10 square chains is 1 acre, a convenient relationship.

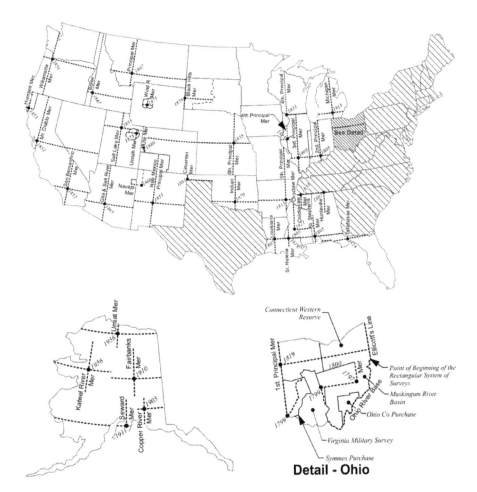

FIGURE 5.1
Initial points of the federal system of rectangular surveys.

A *Principal Meridian* was extended north and south from the initial point following a meridian of longitude, again astronomically determined, on which corner monuments were set every half mile, every 40 chains. Although the initial point and the Baseline are not usually named, it is common for the Principal Meridian to have a name such as, the *Wind River Principal Meridian*, the *Ute Principal Meridian*, or the *6th Principal Meridian*.

These two lines, the Baseline and the Principal Meridian are the fundamental axes and the foundation from which the PLSS was extended across the country. They are by no means abstract historical curiosities. By their monumentation these lines are a real physical presence today. And the Baselines and Principal Meridians are geographical lines, each as nearly as a parallel of latitude and a meridian of longitude as the practical measurement technology of their day could achieve.

However, it should be borne in mind that much of the early surveying in the PLSS was done with a solar compass and a linked chain. The discrepancies between the design and the reality of the monumented corners may be larger than would be expected if the work were done today with modern equipment. Despite that caveat the original work was remarkably good. Whether good or bad, however, there is no question it takes precedence over subsequent retracements or resurveys. The monuments set during the execution of the original survey and the boundary lines they describe are correct and inviolable by law. It is an important principle of the Public Land Survey System and is mentioned, among other places, in a section of a federal statute originally enacted in 1925, 43 Stat. 1144, it says:

> All the corners marked in the surveys returned by the Secretary of the Interior or such agency as he may designate, shall be established as the proper corners of the sections or subdivision of sections they were intended to designate. ... The boundary lines, actually run and marked in the surveys returned by the Secretary of the Interior or such agency as he may designate, shall be established as the proper boundary lines of the sections, or subdivisions, for which they were intended... (BLM 2009, p. 6)

It is possible to think that since most of the public domain is no longer under the jurisdiction of the Federal Government the quoted statute does not necessarily apply to lands in private hands. However, the laws and courts in most states that were created from public domain follow the line that the rules established by Federal Statutes and elaborated in the various Manuals of Instructions will hold sway over the lands in state and private ownership under their jurisdiction.

At the intersection of the Principal Meridian and the Baseline there is an initial point, which is also represented by an actual monument. In fact, in the *Manual of Instructions of 1902* not only stipulated that initial points should be set in conspicuous locations that are visible from a distance, but that they should be perpetuated by indestructible monuments like a copper bolt set in a rock ledge. Monuments perpetuate the initial points, most have been remonumented, but the original monuments of 8 of the 37 initial points are still standing. The Baselines and Principal Meridians that extend in cardinal directions from them often, though not always, terminate at a state line. Once these axes were in place the next step was the creation of *quadrangles*.

Quadrangles

Quadrangles in the PLSS are large rectangular areas bounded by meridians of longitude and parallels of latitude. They are 24 miles on a side by current design, though that has not always been the dimension used. At each initial

FIGURE 5.2
Survey of quadrangles.

point a Principal Meridian on one side, a Baseline on another bound four such quadrangles, and on the other two sides there is a *guide meridian* and a *standard parallel*, as illustrated in Figure 5.2. The lines around a quadrangle are known as *standard lines*.

A guide meridian was intended to follow a meridian of longitude. A standard parallel was intended to follow a parallel of latitude. And they both were established astronomically with corner monuments set every 40 chains, each half mile, just as was done to create the Principal Meridian and the Baseline. Over a broad area of the PLSS there is just one Baseline and one Principal Meridian originating from one initial point, but there are many quadrangles built from them. And so, these standard lines are numbered as you see in Figure 5.2. At a distance of 24 miles north from the Baseline there is the First Standard Parallel North, and 24 miles south from the Baseline there is the First Standard Parallel South. East from the Principal Meridian 24 miles there is the First Guide Meridian East, and to the west 24 miles there is the First Guide Meridian West. This logical system of numbering was and is carried through the whole region governed by a particular Principal Meridian and Baseline forming 24 miles quadrangles bounded by geographic standard lines.

The reason for this approach was explained in the *Manual of Instructions of 1890,*

> Standard parallels shall be established at intervals every 24 miles, north and south of the base line, and guide meridians at intervals of every 24 miles east and west of the Principal Meridian; the object being to confine the errors resulting from the convergence of meridians, and inaccuracies in measurements, within the tracts of lands bounded by the lines so established. (GLO 1890, p. 18)

Since the guide meridians were laid out following astronomically derived meridians of longitude they converged. Even though the First Guide Meridian East may have been 24 miles east of the Principal Meridian at the Baseline, by the time it has run north for 24 miles and intersected the First Standard Parallel North that guided meridian was considerably closer to the Principal Meridian than when it started the trip. And convergence was unavoidable because they both the Principal Meridian and the First Guide Meridian, the Second Guide Meridian, the Third, and so on were all meridians of longitude. Therefore, as described in the 1890 Manual, and illustrated in Figure 5.2, the guide meridians were not then and are not today continuous. They were designed to stop at each standard parallel and the Baseline, so they could be corrected back to their original 24-mile spacing.

At the south end of each of its segments a guide meridian was intended to be 24 miles from the Principal Meridian, or neighboring guide meridian. But at its north end, where it terminated at the standard parallel or Baseline, the spacing between meridians was inevitably less than 24 miles. It was less per the convergence of the meridians. As a consequence that segment of the guide meridian stopped there. A new segment begun, but before that next segment of the guide meridian proceeded north for the next 24 miles it was shifted back to the correct spacing. It was corrected at the standard parallel to be again 24 miles from the Principal Meridian, or its neighboring guide meridian. In this way, it was ensured that the south end of each segment of a guide meridian was then and is today 24 miles east or west from its neighboring guide meridian at its south end. But at the north end of each segment of a guide meridian was and is less than 24 miles from the next guide meridian because of convergence. And this is the reason that standard parallels are also known as *correction lines*. They are the lines at which guide meridians are corrected for convergence.

It is sensible then that the process of actually laying out guide meridians proceeded from the south to north. Corner monuments were set every 40 chains along the way, just as they had been done on the Principal Meridian and the Baseline. When the guide meridian finally intersected the standard parallel a *closing corner* monument was set. A closing corner was set where a surveyed boundary, the guide meridian, intersected a previously established boundary, the standard parallel or Baseline. Any excess or deficiency in the measurement that had accumulated in the work along the guide meridian was placed in the last half mile preceding its intersection with the east–west standard line. So the last half mile at the north end of a segment of a guide meridian may be more or it may be less than 40 chains. More importantly the cardinal direction of the guide meridian was maintained all the way to its intersection with the standard parallel or Baseline. And therefore that intersection could not possibly fall on the already established corner monument on the parallel it was intersecting, because the guide meridian was no longer at the 24-mile spacing it had when it started.

Before the guide meridian was run a *standard corner* had been set on the standard parallel or Baseline 24 miles from the Principal Meridian or guide meridian to serve as a corner of that quadrangle. However, that standard corner could only be the correct corner of the quadrangle to the north, because when the guide meridian was closed on the standard parallel or Baseline convergence caused the intersection of the two lines to fall somewhat west or east of the standard corner. So, in the end there was a standard corner and a closing corner at every corner of a quadrangle and they were some distance from each other. The only exceptions to this are the quadrangle corners set on the Principal Meridian, including the initial point. In every other case there was a closing corner monument set. It was and is the corner for the quadrangles to the south. The corner monument for the quadrangle to the north will usually be a standard corner.

Townships

The quadrangle then became the framework enclosing the land from which 16 *townships* could be created. Townships are the unit of survey of the PLSS. A township is approximately 6 miles on a side and is designed to include 36 sections. The boundaries of a township are intended to follow meridians of longitude and latitudinal lines.

Please recall that as the standard lines around the quadrangle were run corner monuments were set at increments of half a mile, 40 chains. Some of those corner monuments were destined to become *township corners*. These were those set every 6 miles, 480 chains, along the standard lines, which are the boundaries of the quadrangle. From these corners the townships were established.

In the actual surveying of township the *meridional*, north–south, boundaries had precedence. They were surveyed from south to north through the 24-mile block of land, the quadrangle. These north–south township boundary lines were laid out along meridians of longitude as nearly as possible. As the meridional lines were laid out, the corner monuments were set every 40 chains, every half mile. On the other hand, the latitudinal boundaries of the township, the east–west lines, were first run on a trial or *random* line from east to west on most township lines. Random lines were also known as *blank* lines. Corner monuments were only set every 40 chains as the latitudinal line was corrected to a *true* line, usually west to east in these cases.

This random and true line method was used when corner monuments were in place at both ends of a boundary, but the boundary had not yet been actually connected by a surveyed line. It worked this way, a trial line, that is the random line, was created when the surveyor pushed his survey west from the township corner monument he had just established while he was surveying north up the meridian, say, at the northeast corner of a township. His next objective was to survey to the corner monument he knew had already been set 6 miles away on the western boundary of the township, its northwest corner. As he proceeded toward it on the random line he set temporary points every 40 chains. When finally he intersected the western boundary of the township most likely he missed his objective corner. When that happened the distance, north or south that his random line missed the northwest corner was known as the *falling*. The falling was very instructive; it told the surveyor the distance he needed to correct his random line, so it would actually connect the northwest and northeast corner monuments with a true latitudinal boundary. So, the surveyor would return east on the corrected line, this time from the northwest township corner to the northeast township corner. And along the way he would correct the temporary points he had set on the random line and set the actual corner monuments on the true line every 40 chains, except those on the westernmost half mile.

All excess or deficiency of the six-mile measurement was placed in the westernmost half mile. For example, suppose that the surveyor found that instead of 480 chains from the northeast to the northwest corners of the township, he measured 479.80. The design of the PLSS dictated that the deficiency of 0.20 chains is not to be distributed evenly through the 6-mile line. It was and is all absorbed in the final half mile. The result is 11 of the distances between the corner monuments on the true line were 40 chains as expected and the 12th, the westernmost, was 39.80 chains. In this way there were always as many regular sections in a township as possible.

Here is how the procedure was described in the *Manual of Instructions 1902*:

130. Whenever practicable, the township exteriors in a block of land 24 miles square, bounded by standard lines, will be surveyed successively through the block, beginning with those of the southwestern township.

131. The meridional boundaries of the township will have precedence in the order of survey and will be run from south to north on true meridians, with permanent corners at lawful distances; the latitudinal boundaries will be run from east to west on random or trial lines, and corrected back on true lines. The falling of a random, north or south of the township corner to be closed upon, will be carefully measured.

132. When running random lines from east to west, temporary corners will be set at intervals of 40.00 chains, and proper permanent corners will be established upon the true line, corrected back in accordance with these instructions, thereby throwing the excess and deficiency against the west boundary of the township as required by law. (GLO 1902, p. 57)

The order of the running the lines to create the townships out of the quadrangle is illustrated in Figure 5.3. It began at the southeast corner of the southwestern township. The first line run was the eastern boundary of that township that was terminated with the setting of the northeast township corner. It is a meridional boundary. These north–south township lines are also known as *range lines*. The next line was a random line along the northern latitudinal boundary of the first *tier* of townships. After the random line was corrected for the falling it was returned on a true line back to the northeast township corner. Then the range line was continued northward, the next township corner was set, and random and true lines were run to the township corner on the standard line. The procedure was repeated on the next township boundaries north, and finally the range line was closed on the northern parallel of the quadrangle. That closing corner, labeled CC in Figure 5.3 was established in much the same way as the northeast and northwest corners of the quadrangle described earlier. And once again the excess and deficiency in the measurement of the range line was placed here in the northernmost half mile, and so it might be more or less than 40 chains.

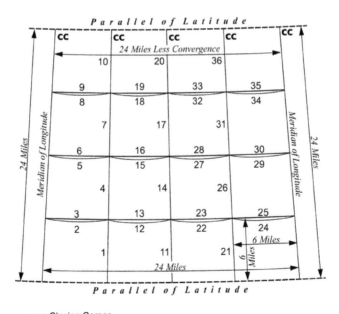

cc = Closing Corner

FIGURE 5.3
The order of running the lines in the subdivision of a quadrangle into townships.

The surveying of the township boundaries continued and the pattern was repeated for the next range line. It was also begun at its south end. Corner monuments were set every 40 chains for 6 miles as the meridional line was surveyed northward. Once again the surveyor ran a random line west, but this time the objective corner was not on a standard line, but the recently set northeast township corner. The falling was noted and the line returned on the true line, as corner monuments were set every 40 chains. These procedures were continued until the range line was closed on the northern quadrangle boundary. The last range line in the quadrangle required a variation. It was begun on its south end, like the others. After being run north for 6 miles a random line and a true line were run west, like the others. But after the true line was returned, another random line was run from the township corner east to the eastern boundary of the quadrangle and returned on the true line. Here too the excess and deficiency of the 6-mile measurement was still absorbed in the westernmost 40 chains.

When the subdivision of a quadrangle into 4 tiers and 4 ranges of townships were done according to the design of the PLSS 16 townships whose boundaries were very nearly meridians of longitude and parallels of latitude were created. The inevitable errors in the east–west distance measurements were taken up in the last half mile on the west side of each township. The errors in the north–south measurements were thrown into the last half mile on the north side of the northern tier of townships, just south of the north boundary of the quadrangle. And all 4 boundaries of all 16 townships had corner monuments in place about every 40 chains.

In fact, such clarity was seldom achieved. There were almost always some extenuating circumstances that interrupted the orderly execution of the standard PLSS plan. Quadrangles that contained large or navigable rivers or other large bodies of water could not be surveyed so neatly. Quadrangles that contained previous bona fide rights in land that was outside the purview of PLSS, such as patented mineral claims or Spanish land grants required that the orderly pattern of running lines be interrupted. In these cases the normal lines were closed upon the boundaries of the senior property rights or the limits of the water as defined by a *meander* line, more about that in a moment. In fact, there were myriad difficulties that frequently prevented straightforward subdivision of a quadrangle into townships. Nevertheless, the fundamental plan was most often carried out much as described. And when difficulties did arise, the surveyors involved did their utmost to use procedures, however unorthodox, that created a subdivision that fit the design.

Sections

The township then became the framework around the land from which 36 sections could be created. Since 1796 the numbering of the sections has begun in the northeast corner of the township. It has been sequential from east to west in the first tier of sections, 1 through 6. Section 7 has been south from section 6 and the second tier has been sequential from west to east, 7 through 12. Section 13 has been south from section 12 with the third tier sequential from east to west again. This zigzag has continued until section 36 in the southeast corner of the township. This pattern is *boustrophedonic*, that is, it alternates from right to left and then left to right. It is said that the system harks back to an ancient mode of writing and the path an ox takes plowing a field. Whatever the origin of the scheme, the same basic plan is also used on the numbering of *fractional lots* within the sections, more about that later.

Sections are the unit of subdivision of the PLSS. They are approximately 1 mile on a side and contain approximately 640 acres. Since 1890 the boundaries of a section have not been intended to follow meridians of longitude and parallels of latitude. Section lines are intended to be parallel to the *governing boundaries* of the township. The governing boundaries are the east and the south township boundaries, unless those township boundaries are defective.

Even though the procedure for the present method of subdividing a township into sections was fairly well established by 1855 (Figure 5.4), here is a description of the procedure from the *Manual of Instructions of 1930*.

> The subdivisional survey will be commenced at the corner of sections 35 and 36, on the south boundary of the township, and the line between sections 35 and 36 will be run parallel to the east boundary of the township ... establishing the corner of sections 25, 26, 35 and 36. From the last named corner, a random line will be run eastward ... parallel to the south boundary of section 36 to its intersection with the east boundary of the township, placing 40 chains from the beginning a post for temporary quarter-section corner.... If the random line intersects said township boundary to the north or south of said corner, the falling will be calculated and the true line ... established.... The meridional section line will be continued on the same plan, likewise the latitudinal section lines. After having established the west and north boundaries of section 12, the line between sections 1 and 2 will be projected northward on a random line parallel to the east boundary of the township ... to its intersection with the north boundary of the township.... If ... said random line intersects the north boundary of the township to the east or west of the corner of sections 1 and 2 the falling will be carefully measured and from the data this obtained the true course returned course will be calculated. (GLO 1930, p. 181)

In other words, when a township was subdivided the eastern range of sections was first in line. The work began at the southwest corner of section 36 and the line run parallel with the eastern boundary of the township. A corner monument halfway, at 40 chains, was called a *quarter* corner, and a corner monument set at 80 chains from the beginning was the section corner. Quarter corners are so named because connecting these monuments with straight cardinal lines across a section divides it into quarters.

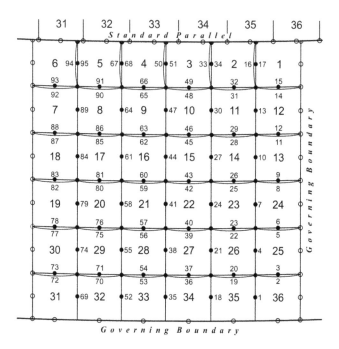

FIGURE 5.4
The order of running the lines for the subdivision of a township into sections.

Next the boundary between sections 36 and 25, a latitudinal line, was run on a random line, headed for the objective corner at the northeast corner of section 36. Once the falling was discovered, the line was corrected and run back, with the final quarter corner monument set halfway. This procedure was repeated all the way up to the meridional line between sections 1 and 2. At this point there were two possibilities.

If the northern line of the township was not a standard parallel, the north–south line was run on a random line toward the objective corner that had been set when the north line of the township had been laid out. A temporary quarter corner was set at 40 chains from the south, with the excess and deficiency of the measurements thrown into the last half mile. Once the falling was found to the objective corner the true line returned southward and the temporary quarter corner replaced with a permanent monument, still there were 40 chains north of the section corner to the south.

On the other hand, if the northern line of the township was a standard parallel the north–south line was still parallel with the east township boundary. However, in this case there was not an objective corner. Please recall that in this circumstance the monuments that had been set when the north line of the township, the standard parallel, was laid out were only applicable to the townships north of the standard parallel, because of convergence of meridians. Nevertheless during the subdivision of the township into sections, a quarter corner was set at 40 chains from the south, with the excess and deficiency of the measurements thrown into the last half mile. A closing corner was set at the intersection of the meridional section line with the standard parallel, and the distance between the closing corner and the previously set corners on each side of it were recorded.

Using this plan a township could be fully subdivided. A deviation from the procedure occurred when the last meridional line was run starting between sections 31 and 32. The random and true line method was used to establish the latitudinal lines to both the east and west.

The final result of the work was a township that contained 36 sections. Working with sections today it is clear that most are nearly a mile square. But those along the northern and western boundaries of the township are not so regular, nor were they intended to be. The design of the system itself pushes the distortion from convergence and the accumulated measuring errors into the last half mile of the tier of sections along the north and the range of sections along the west. It is there that the excesses and deficiencies are absorbed, but only after as many regular sections as possible have been carved out of the township.

Creating as many regular sections as possible has been an objective of the PLSS from the beginning, but the rectangular system has had to be flexible. The surface of the Earth does not easily accommodate such a regular geometry. Applying the plan to lands adjoining rivers and lakes, areas where there may be an incomplete south or east township boundary has called for ingenuity. These *fractional* townships often made it impossible to follow the general rules laid down in the Manuals of Instructions. Joining the townships and sections from one initial point region with those from another and the irregularity of defective township exteriors also tends to disrupt the design.

The Subdivision of Sections

As mentioned earlier straight lines run from the quarter section corner monuments that stand in the middle of the four section lines dividing the section into quarters. This method was stipulated in the *Act of February 11, 1805*. The intersection of these lines is considered the center of the section, or center quarter corner, and is a corner common to the four-quarter sections. Government surveyors set the quarter section corner monuments when they created the sections inside the township, but generally they did not subdivide the sections. Running the lines not done by the government surveyors, including the subdivision of sections has always been left to the local surveyors. Other Acts of Congress in the early nineteenth century provided definitions of the half-quarter and quarter-quarter portions of a section, 1820 and 1832 respectively. But actually creating those parts on the ground was again left to local surveyors.

These quarter and half divisions are known as *aliquot* parts of a section. A division of something is aliquot when the divisor is an integer and there is no remainder. But in practical terms this has meant that an aliquot part of a section in the PLSS is a half or a quarter of the previously larger subdivision. These are its legal subdivisions. The PLSS does not recognize other divisions such as thirds of fifths. For example, quarter sections can be further divided into halves and quarters and they are legal subdivisions as well. The quarter-quarter section, that is, a quarter of a quarter section, is the smallest legal subdivision per the *Manual of Instruction of 2009*. The 2009 Manual is the most current version of that long series. In other words, a normal section contains approximately 640 acres. The typical aliquot parts of such a section are four. Half sections contain approximately 320 acres. Quarter sections contain approximately 160 acres. Half-quarter sections contain approximately 80 acres. Quarter-quarter sections contain approximately 40 acres.

Township Plats

When the notes of the original surveys of the townships in the Public Lands were returned to the GLO, and later the BLM, plats were developed from the work. These plats represent the township included in the survey. They show the direction and length of each line and their relation to the adjoining surveys. They include some indication of the relief, the boundaries, descriptions, and area of each subdivision of the sections. As mentioned earlier the government did not and will not convey lands in the public domain to others until the survey has been done and the official plat has been filed and approved. This is true in part because the deed, the patent, which does eventually grant the ownership of the land, does so by direct reference to the township plat. It is an integral part of the transaction and binds the parties to the specific dimensions of the land as shown on the township plat. Therefore, the orientation and size of the parts of the sections on the township plat are significant. Significant too then is the creation of the lines on the township plat by *protraction*.

Protraction in this context refers to drawing lines on the township plat that were not actually run or monumented on the ground during the official subdivision of the land. For example, the lines that divide the sections into quarters, the quarter lines, are shown on the township plat as dashed lines. They are protracted, because even though the quarter corner monuments were set, the official government surveyors never ran the quarter lines themselves. Rules were drafted to guide the local surveyors that would actually run those lines eventually. Rules that stipulated that they must be straight from quarter corner to quarter corner, and their intersection is the center of the section, but the government was not responsible for running the lines themselves. They drew them on the plat. The same may be said of the lines that divided the quarter sections into quarters, the quarter-quarter sections. They are also protracted lines and were not run on the ground by the GLO or BLM, except under special instructions.

Therefore, it follows that for all practical purposes the aliquot parts of a section, that means virtually all the lines and the areas shown inside a section on an official township plat, were created by protraction. They were drawn on the plat, not run on the ground. And that includes the boundaries of another legal subdivision of a section that is not aliquot at all. It is known as a fractional lot.

Fractional Lots

Lotting occurs in fractional sections and the definition of fractional section has grown a bit over the years. GLO Commissioner George Graham wrote one of the earliest definitions of the term to George Davis in 1826,

> It is here proper to premise that the technical meaning of 'fractional section' is a tract of land not bounded by sectional lines on all sides in consequence of the intervention of a navigable stream or some other boundary recognized by law and containing a less quantity than six hundred & forty acre. (White 1983, p. 83)

However, the definition has expanded somewhat over the years. It still applies to the sections the Commissioner mentioned, but it also includes the sections on the north and west of a normal township. Please recall that the excess and deficiency of the measurements made during the subdivision of a quadrangle into townships and the subdivision of a township into sections is absorbed in the last half mile along the northern and western boundaries. And that means that, by design, the last half of those sections is something more or less than 40 chains, 2640 ft. One of the principles of the PLSS is always to create as many regular aliquot parts as possible. In practice this means that half of that last half mile of a section on the northern and western boundaries of a township can be held to 20 chains. Doing that creates the last regular aliquot part before hitting the north or west boundary of the township. But the remainder of that distance is then to become a side of a parcel that cannot be described as an aliquot part. It cannot be an aliquot part because it is not, nor can it be, a half or a quarter of the previously larger subdivision. Therefore, it is known as a fractional lot.

All along the north and the west boundary of a normal township one will find one row, or more, of these fractional lots. The sections bordering the north and west boundaries of the township, except section 6, are normally subdivided into two regular quarter-sections, two regular half-quarter sections, and four fractional lots. In section 6, the subdivision will show one regular quarter-section, two regular half-quarter sections, one regular quarter-quarter section, and seven fractional lots. This is the result of the plan of subdivision.

That is by no means the only place that fractional lots are found. Any section whose boundaries are incomplete may be a fractional section, a section invaded by a body of water, for example. In such a case a meander line segregates the water and the section or sections around it are fractional.

A meander line is a line run along the mean high water mark of a permanent natural body of water. The meander line generally follows the curves, twists, and turns of the shore. It is run to determine the area of the land in a fractional section, after the area covered by the water is excluded. Since its inception the PLSS has been concerned with transferring land from the public domain to others, and therefore the area of usable property contained in the aliquot parts of sections has been a matter of keen interest. However, a navigable river is not a part of the public domain. It must forever remain a common public highway and is excluded from the system. Under current instructions a body of water is meandered if it is a navigable river, is a stream more than 3 chains wide, or a lake that is larger than 50 acres. It is important to note that the meander line around the water is not actually a property boundary, that is, a line actually defined by the water itself; nevertheless, closing corners are set on meander lines. When a standard line, township line, section line, or aliquot line intersects a meander line, a *meander corner* is set to perpetuate the intersections. Where that happens fractional lots spring up. They are protracted against meander lines.

The same thing happens when a more recent PLSS survey collides with a former official government survey. For example, it is inevitable that PLSS surveys originating at one initial point will grow outward and eventually meet the PLSS surveys that originate from another initial point. When that occurs, the lines and areas of the two systems will certainly not match. Fractional lots are protracted where they join. When the PLSS survey encounters boundaries around property that existed before the rectangular system got there, fractional lots are protracted again. This is the case when lines are closed on a reservation, national park, mining claim, previous land grants, or other bona fide private claim. In each of these instances the guiding principle is that fractional sections should be subdivided in a way that produces as many aliquot parts as possible.

Fractional lots are numbered. The system of their numbering follows the same model as that used to number sections. It is a zigzag pattern that begins at the most northeasterly lot, which is Lot 1. Then continuing within the same section, the next lot to the west is Lot 2. It is important to note that the next lot is not necessarily immediately adjacent to Lot 1. The general plan is that Lot 1 through Lot 4 are numbered east to west sequentially as shown in Figure 5.5. The numbering scheme then drops down to the next tier of lots

FIGURE 5.5
Fractional lots.

south from Lot 4 to Lot 5. Then the plan would be to sequentially number lots from west to east, Lot 5 through Lot 8, drop south to the next tier, and number the lots east to west, Lot 9 through Lot 12. Finally, the southern-most tier would be numbered west to east Lot 13 through Lot 16. However, there are not necessarily fractional lots in each tier and aliquot parts are not numbered. Therefore, in a typical section 6 the lots on the northern tier are numbered east to west 1 through 4, and Lot 5 is south of Lot 4 accord-ing to plan. But from Lot 5 east there are usually aliquot parts, no lots. Lot 6 then is numbered south of Lot 5 because it is simply the next fractional lot encountered by the zigzag numbering system. In other words the scheme can be imagined to proceed east from Lot 5 seeking lots to number, but none are there. It then drops south to the next tier, still none. It then proceeds east to west, and finally along the western boundary encounters a fractional lot then becomes Lot 6. The same thing happens between Lot 6 and Lot 7.

The particular Lot 6 that is shown in Figure 5.5 is crossed by a navigable river. The northernmost tier of lots is numbered 1 through 4 from east to west. Then the second tier of lots is numbered west to east starting with Lot 5, which is directly south of Lot 4. East of Lot 5 is an aliquot part, a quarter-quarter. East of that part is the fractional Lot 6. It cannot be an aliquot part because a corner of it is inside the meander line that follows the northern bank of the river. East of Lot 6 is Lot 7 and it is also partially in the river. The east boundary of Lot 7 is the section line, so the numbering of the fractional lots drops south the next tier and proceeds east to west again. South of Lot 7 area is in the river, but directly west of the river is Lot 8. West of Lot 8 is an aliquot part followed by Lot 9 on the western boundary of the section. The numbering scheme drops south to the next tier and Lot 10. It then proceeds from west to east, jumps across the aliquot part, and then Lot 11 is num-bered. The last lot numbered is Lot 12 in the southeast corner of the section. It is interesting to note that Lot 12 includes a small area that could be in the tier of lots to its north. However, one of the rules of the creation of fractional lots specifies that very small lots will not be protracted, but such an area will instead be included into adjoining lots, as is illustrated here.

Naming Aliquot Parts and Corners

As was just mentioned, aliquot parts of a section are not numbered. They are named.

Some aspects of their names are intuitive. For example, 160-acre portion of a section north and east of its center can simply be called the north–east quar-ter. It can also be abbreviated as NE¼. Describing any of the other quarters of a section can be done similarly, such as NW¼, SW¼, and SE¼. Describing quarter-quarters of a section is done with a combination of abbreviations. For example, the 40-acre portion of the NW¼, that is, in the most northwesterly corner of the section is known as NW¼NW¼. This description can be read, the northwest quarter of the northwest quarter. The other quarter-quarters in the NW¼ could be described with the same system as: NE¼NW¼, SE¼NW¼, and SW¼NW¼. There are a couple of important keys to such an abbreviation system. First, the smaller tract is always first, meaning it is on the left. The larger tract is on the right. Second, there is no comma between the two elements. A comma would mean, *and*, in this context. In other words NE¼ NW¼ would mean the north–east quarter and the north–west quarter when the intention of NE¼NW¼ is to indicate the north–east quarter of the north–west quarter. The general structure of the system is that the term on the left is a portion of the term on the right. The same ideas can be used to describe the east half of the south–east quarter. It could be described as E½SE¼. Fractional lots can be described using their numbers: Lot 1, Lot 2.

The abbreviated descriptions of Lots and aliquot parts do not fully define them however. Writing a complete and correct description requires the addition of the number of the section, the township, the range, and the Principal Meridian. For example, the S½SE¼ Section 36, Township 3 South, Range 68 West of the 6th Principal Meridian is complete, concise, and unambiguous. It can only refer to one parcel of land.

The naming of section corners in PLSS is also straightforward. The north–east, north–west, south–east, and south–west corners of a section are simply abbreviated as: NE, NW, SE, and SW respectively. The quarter corners too are simply stated. The corners at the midpoints on the section lines are the north quarter corner, east quarter corner, south quarter corner, and west quarter corner that are abbreviated as: N¼, E¼, S¼, and W¼ respectively. The center quarter, which is at the intersection of the quarter section lines, is abbreviated as C¼.

It is important to mention that another common name for a quarter-quarter of a section is a sixteenth. Therefore, corners of quarter-quarters are often called sixteenth corners, such as the corners in the middle of quarter sections where the quarter-quarter lines intersect.

For example, as shown in Figure 5.6 the corner in the centers of the NE¼ of the section common to the four quarter-quarters is called the NE¹⁄₁₆. The other ¹⁄₁₆th corners in the centers of the quarter sections are also named for the quarter in which they are set. They are called NW¹⁄₁₆, SE¹⁄₁₆, and SW¹⁄₁₆. The abbreviations for the sixteenth corners on the quarter lines of the section refer first to the fact that the lines on which they fall are the centerlines of the section. For example, the corner on the N–S centerline between the NE¼ and the NW¼ is called C-N¹⁄₁₆. The other center quarters are abbreviated in the same pattern: C-E¹⁄₁₆, C-S¹⁄₁₆, and C-W¹⁄₁₆. Finally, the ¹⁄₁₆th corners on the sections lines themselves follow a similar scheme. Their abbreviations refer to the section line on which they stand first, and then whether they are east or west from the quarter corner on the same section line. Therefore, on the north line of the section there are the N-E¹⁄₁₆ and N-W¹⁄₁₆ corners. On the east

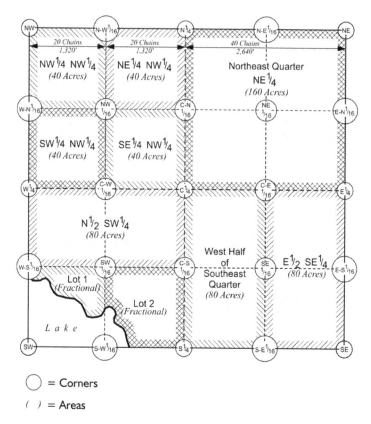

FIGURE 5.6
A section.

section line there are two ¹⁄₁₆th corners: E-N ¹⁄₁₆ and E-S ¹⁄₁₆. The abbreviations applied to the ¹⁄₁₆th corners on the south section line and the west section lines are created with similar logic.

This upper portion of Figure 5.7 shows the normal subdivision of sections 5, 6, 7, and 8 as per the design of PLSS, and the lower portion shows a portion of a township in Wyoming, T18N R89W of the 6th Principal Meridian as it actually stands. Portions of the west boundary, north boundary, south boundary, and the subdivisional lines were the subject of a dependent resurvey by a Supervisory Cadastral Surveyor in 1963 under special instructions that restored some lost corners to their true original locations. The work was concerned with protecting the property rights that had arisen from the original survey as well. For example, the fractional lots are shown as given on the original plat of the township, which was approved in 1884. In other

FIGURE 5.7
The plan and the actual.

words, the surveyor in 1963 did everything he could to reconstruct the positions of the lines and corners, as they had existed after the original survey of 1884. And this is consistent with the mandate for local surveyors throughout the states that were constructed from the public domain. When working in the PLSS their objective is always to preserve or reestablish the original locations of the lines and corners if at all possible.

However, by comparing the two parts of Figure 5.7 one can see that there is often a difference between the pure design of the PLSS and the actual section lines and corners as the original surveyor laid them out on the ground. Nevertheless, it is certain that the monuments set and lines run on the ground by that original surveyor are conclusive. The corner monuments originally set and the section lines originally run are inviolable. Here are some of the fundamental rules, which control in the PLSS system and underline the point:

> First: That the boundaries of the public lands, when approved and accepted are unchangeable.
>
> Second: That the original township, section and quarter-section corners must stand as the true corners which they were intended to represent...
>
> Third: That the quarter-quarter-section corners not established in the original survey shall be placed on the line connecting the section and quarter-section corners, and midway between them, except on the last half mile of section lines closing on the north and west boundaries of the township, or on the lines between fractional or irregular sections.
>
> Fourth: That the center lines of a section are to be straight, running from the quarter-section corner on one boundary to the corresponding corner on the opposite boundary.
>
> Fifth: That in a fractional section where no opposite corresponding quarter-section corner has been or can be established, the center of such section must be run from the proper quarter-section corner as nearly in a cardinal direction to the meander line, reservation or other boundary of such fractional section, as due parallelism will permit.
>
> Sixth: That lost or obliterated corners of the approved surveys must be restored to their original locations whenever this is possible. (BLM 1979, p. 6)

It is interesting to note the practical effect of these rules. Section lines and quarter-section lines are correct as they were originally run, even though typical regular section lines are not 1 mile long, and quarter section lines are not usually a half-mile long. It is usual that the quarter corners of a section are not actually midway between section corners, or are they in line with section corners. Nevertheless these monuments are the corners they are intended to represent, exactly. Despite the fact that section lines most often do not have a cardinal bearing and the legal center of the section is not the geometric center of the section, the lines as originally surveyed are right. Even though it may be that none of the angles at the section corners is 90° and the opposite sides of the section are not parallel with each other, the lines as originally surveyed as conclusive. Few sections actually contain 640 acres; in fact, none of its other legal subdivisions usually contain their theoretical acreages. And the actual dimensions of a section measured today seldom match the original record dimensions. But despite all these deviations from the plan the section is absolutely correct as it was originally laid out on the ground. This overarching principle was codified long ago, by the *Act of February 11, 1805*. Corners and lines established during the original government survey are unchangeable.

That is not to say that there are no limits to how far a section, township, or quadrangle may depart from the ideal in the PLSS. The general rectangular limits are

> (1) For alinement [*sic*], the section's boundaries must not exceed 21′ from cardinal in any part, nor many opposite (regular) boundaries of a section vary more than 21′.

> (2) For measurement, the distance between regular corners is to be normal according to the plan of survey, with certain allowable adjustments not to exceed 25 links in 40 chains.
> Township exteriors, or portions of exteriors, are considered defective when they do not qualify within the above limits. It is also necessary, in order to subdivide a township regularly, to set a third limit, as follows:

> (3) For position, the corresponding section corners upon the opposite boundaries of the township are to be so located that they may be connected by true lines which will not deviate more than 21′ from cardinal. (BLM 2009, p. 70)

However, it should be noted that these rules apply to work done under the current *Manual of Surveying Instructions 2009*. If the surveying were done under other older instructions, and most of the public domain was subdivided long ago, then the work is not necessarily bound by these limits. Along that line it is important to remember that the rules that governed the original establishment of the corners and lines in the PLSS have not always been exactly as they are today. For example, there was no accommodation of the convergence of meridians and sections that were numbered sequentially from south to north in the original seven ranges of sections laid out in 1785 in Ohio. Nevertheless, no attempt would be made to alter those sections to harmonize with rules in place today. For example, prior to 1855 the distance between standard parallels was often 30 miles or more. In some cases they were as much as 60 miles apart. When working with townships and sections that were originally established in that era it is important to recognize that they are correct, even though the current instructions place the distance between standard parallels at 24 miles. There are many differences between current guidelines and those of the past.

Here is an example of the evolution of section and quarter section corners. It is just one instance of official PLSS surveying that was done contrary to the current plan, but was nevertheless absolutely correct at the time. In the *Act of May 18, 1796* there was an economy measure. Only every other township was actually subdivided into sections. And in those townships that were subdivided, only every other section line was run on the ground. In other words, in half of the townships the section lines were run every 2 miles, and corners were marked on those lines every mile.

The *Act of May 10, 1800* was passed to amend the *Act of May 18, 1796*. It had proven impractical to expect the purchasers of sections to set the many corner monuments that the government surveyors had not established. They were stubbing them in by the easiest methods and that was producing large errors. Therefore, in 1800 Congress mandated that section lines would hence-forth be run every mile, not every 2 miles. Corner monuments would be set each mile on the north–south lines and every half mile on the east–west lines. Thereby the system evolved from setting no quarter corners, to setting only those on latitudinal lines. Also, the excesses and deficiencies of the measure-ments would be placed in the last half mile on the north and west boundary of the township. It took another five years, under the *Act of February 11, 1805*, for the quarter corners on the meridional section lines to be called for.

It follows that when working with sections that were established before 1805 it might be valuable to recall that all the quarter corners were not set in the original survey. This is just one of the many discrepancies from the cur-rent instructions that guide the PLSS today.

Another example, the *Act of June 1, 1796* mandated the establishment of townships that were, and still are 5 miles on a side instead of 6 miles. Congress had segregated a portion of land known as the U.S. Military Reserve to satisfy the military warrants it had issued to pay Revolutionary War soldiers. These warrants were in multiples of 100 acres and 6-mile townships contain 23,040 acres, not evenly divisible by 100. The 5-mile townships contained 16,000 acres and were much more convenient to the purpose. And that is not the only region in which special instructions stipulated deviations from the normal plan.

There are two eminent books that are reliable guides in this regard. *The Manual of Surveying Instructions 2009* and *A History of the Rectangular Survey System* both are published by the United States Government Printing Office. The former book was compiled under the auspices of the Bureau of Land Management, Department of the Interior and is a clear, concise manual that sets out the guidelines for the current Public Land Survey System plan. The latter book was written by C.A. White and is also published by the U.S. Government Printing Office. It is a valuable resource that explains the rules and regulations that governed the PLSS from the beginning till today. It also includes reproductions of many of the former manuals and instructions issued by Surveyors General in the past. Together these books provide practical and valuable insights to both the past and the present PLSS.

Finally, the Public Land Surveying System was the most extensive surveying project ever undertaken. It has been extraordinarily successful. It has evolved. The rules governing its execution have been practical. Still, there are large portions of mountains, wilderness areas, and deserts that have not been surveyed in the PLSS. Unfortunately, unofficial township and section lines are sometimes drawn over and across these areas to provide the continuation of the system where it has never actually been applied by government survey. As has been described here the general PLSS plan does not necessarily correspond exactly with the actual corners and lines on the ground, but it is best to recall that it is always those actual corners and lines as originally surveyed on the ground that are correct and conclusive. Where no such corners and lines were ever run the PLSS system is not in place at all.

References

Bureau of Land Management. *Restoration of Lost or Obliterated Corners and Subdivision of Sections.* Washington, DC: U.S. Government Printing Office, 1979.

Bureau of Land Management. *Manual of Surveying Instructions 2009.* Washington, DC: U.S. Government Printing Office, 2009.

General Land Office. *Instructions of the Commissioner of the General Land Office to the Surveyors General of the United States Relative to the Survey of the Public Lands and Private Land Claims, May 3, 1881.* Washington, DC: U.S. Government Printing Office, 1881.

General Land Office. *Manual of Surveying Instructions for the Survey of Public Lands of the United States and Private Land Claims, January 1, 1890.* Washington, DC: U.S. Government Printing Office, 1890.

General Land Office. *Manual of Surveying Instructions for the Survey of Public Lands of the United States and Private Land Claims, January 1, 1902.* Washington, DC: U.S. Government Printing Office, 1902.

General Land Office. *Manual of Surveying Instructions for the Survey of Public Lands of the United States and Private Land Claims, 1930.* Washington, DC: U.S. Government Printing Office, 1930.

White, C. A. *A History of the Rectangular Survey System.* Washington, DC: U.S. Government Printing Office, 1983.

Exercises

1. Which of the following general statements about initial points in the PLSS is incorrect?
   a. Most of the initial points were already in place when they were first mentioned in the Manual of Instructions.

b. A latitude and longitude were chosen and an initial point set at that coordinate before the PLSS survey was commenced in a region.

c. The geographic coordinates of the initial points were determined astronomically.

d. Several of the initial points still retain their original monumentation.

Explanation:

Initial points are the origins of the PLSS. There are 32 in the conterminous United States, 5 in Alaska, and several of them do still have their original monuments. When they were first mentioned in the *Manual of Instructions* in 1881 most of the 37 points were already in place. However, the criteria for setting initial points never included the idea that they needed to be set at a particular latitude and longitude. Their latitudes and longitudes were determined astronomically and recorded, but they were not placed at a previously determined coordinate.

Answer is (b)

2. Which of the following states has no initial point within its borders?
   a. Wyoming
   b. Oregon
   c. Iowa
   d. Montana

Explanation:

PLSS surveys in Iowa are referred to the 5th Principal Meridian, which is in Arkansas. However, there is no initial point in Iowa. There are several other states in which there is no initial point.

Answer is (c)

3. Which of the following categories of corner monuments was not and is not set during an official government survey, unless it is done under special instructions?
   a. Section corners
   b. Quarter section corners

    c. Township corners

    d. Quarter-quarter corners

Explanation:

Quarter-quarter corners are not set under current instructions and they have not been previously mandated. Special instructions can be issued that would stipulate that such corners be monumented during the official survey.

Answer is (d)

4. Under normal conditions and by current instructions, which of the following statements is true concerning both the subdivision of a quadrangle into townships and the subdivision of a township into sections?

    a. The survey begins on the south line, and corner monuments are set every 40 chains on parallels of latitude and meridians of longitude.

    b. The official units of the survey are chains and corner monuments as set by the original surveyor are inviolable.

    c. The random and true line procedure is used on the latitudinal lines and the east and south lines are the governing boundaries.

    d. The survey begins on the south line, corner monuments are set every 40 chains and the lines are intended to be parallel.

Explanation:

Although it is true that the subdivision of a quadrangle into townships and the subdivision of a township into sections both begin on the south line, only the boundaries of townships are intended to be geographical lines. The section lines in a township are intended to be parallel with the south and east boundaries of the township. Corner monuments are typically set every 40 chains, and chains are the official unit of the PLSS. Original monuments and lines established during the original survey are inviolable.

Answer is (b)

5. Which of the following have remained unchanged since the beginning of the PLSS?

    a. The pattern of numbering sections in a township.

    b. The method of subdividing a township into sections.

c. The size of a township.

d. The principle that land in the public domain must be surveyed before it is sold.

Explanation:

Until 1796 the numbering of sections in a township was done in a north–south scheme, since then the east–west zigzag pattern used today has been unchanged. Also in 1796 only every other township was actually subdivided into sections. And in the half of the townships that were subdivided, section lines were run only every 2 miles and corners were marked on those lines every mile. Under the *Act of June 1, 1796* 5-mile townships were surveyed in the land set aside as the U.S. Military Reserve to satisfy the military warrants Congress had issued to pay Revolutionary War soldiers. However, the principle that land must be surveyed before it is sold has been fairly consistent in the history of the PLSS.

Answer is (d)

6. Under normal conditions and by current instructions, which section's boundaries are completed last in the subdivision of a regular township?

a. Section 1

b. Section 6

c. Section 31

d. Section 36

Explanation:

Under normal circumstances the subdivision of a township into sections begins at the southwestern corner of Section 36 and continues through the eastern range of sections.

That work is followed consecutively in the west by each of the six ranges of sections in the township. Therefore, the last section completed is Section 6 in the northwestern corner of the township.

Answer is (b)

7. Which of the following is most probably an original corner monument that was approximately 72 miles west and 95 miles north from an initial point?

a. It is a closing corner at the intersection of the Third Guide Meridian West and the Fourth Standard Parallel North.

b. It is a standard corner on the Third Guide Meridian West at the southwest corner of Section 6, Township 16 North, and Range 12 West.

c. It is a standard corner on the Fourth Guide Meridian West at the southwest corner of Section 6, Township 12 North, and Range 15 West.

d. It is a closing corner at the intersection of the Third Standard Parallel North and the Fourth Guide Meridian West.

Explanation:

There are 24 miles between guide meridians. Therefore, the distance from the Principal Meridian to the Third Guide Meridian West is 72 miles. The corner in the question is on the Third Guide Meridian West. There are also 24 miles between standard parallels. Therefore, the distance from the Baseline to the Fourth Standard Parallel North is 96 miles. The corner in the question cannot be on the Fourth Standard Parallel North. However, it is in the township whose northwest corner is at the intersection of the Third Guide Meridian West and the Fourth Standard Parallel North. That is, Township 16 North and Range 12 West. It is the corner in that township 1 mile south of the Fourth Standard Parallel North. It is the southwest corner of Section 6.

Answer is (b)

8. What aliquot part of a section has the following corners: C¼, NW$\frac{1}{16}$, C-N$\frac{1}{16}$, and C-W$\frac{1}{16}$?

   a. E½ SE¼
   b. NW¼
   c. NW¼ NW¼
   d. SE¼ NW¼

Explanation:

The 40-acre aliquot part known as the SE¼ of the NW¼ has the corners described.

Answer is (d)

9. Which of the following statements about the difference between a quarter-quarter of a section and a fractional lot is correct?

   a. A quarter-quarter of a section contains 40 acres, a fractional lot does not.

   b. A quarter-quarter of a section is an aliquot part, a fractional lot is not.

   c. A quarter-quarter of a section is numbered, a fractional lot is not.

   d. The corners of a quarter-quarter section are set during the original government survey; the corners of a fractional lot are not.

Explanation:

A quarter-quarter of a section does not necessarily contain 40 acres, and it is possible for a fractional lot to contain 40 acres and still not be an aliquot part. A fractional lot is a subdivision of a section that is not a half or a quarter of the previously larger part of a section. On the other hand, a quarter-quarter of a section is an aliquot part. A fractional lot is numbered; a quarter-quarter of a section is not. Finally, neither the corners of the quarter-quarter of a section nor the corners of a fractional lot are monumented during the original survey unless such work is specified by special instructions.

Answer is (b)

10. When a standard, township, or section line intersects a meander line along the high water mark of a navigable body of water what sort of a corner is established?

    a. A section corner

    b. A closing corner

    c. A meander corner

    d. A standard corner

Explanation:

Meander lines do not describe legal boundaries. They are a method used to set off certain bodies of water from the PLSS. Meander corners are established where the meander line along the mean high water is intersected by a standard, township, or section line.

Answer is (c)

Index

Note: Page numbers followed by f and t refer to figures and tables, respectively.

Milton Keynes UK
Ingram Content Group UK Ltd.
UKHW051015071024
449327UK00012B/268